Deep Learning for Computer Architects

Synthesis Lectures on Computer Architecture

Editor
Margaret Martonosi, *Princeton University*

Founding Editor Emeritus
Mark D. Hill, *University of Wisconsin, Madison*

Synthesis Lectures on Computer Architecture publishes 50- to 100-page publications on topics pertaining to the science and art of designing, analyzing, selecting and interconnecting hardware components to create computers that meet functional, performance and cost goals. The scope will largely follow the purview of premier computer architecture conferences, such as ISCA, HPCA, MICRO, and ASPLOS.

Deep Learning for Computer Architects
Brandon Reagen, Robert Adolf, Paul Whatmough, Gu-Yeon Wei, and David Brooks
2017

On-Chip Networks, Second Edition
Natalie Enright Jerger, Tushar Krishna, and Li-Shiuan Peh
2017

Space-Time Computing with Temporal Neural Networks
James E. Smith
2017

Hardware and Software Support for Virtualization
Edouard Bugnion, Jason Nieh, and Dan Tsafrir
2017

Datacenter Design and Management: A Computer Architect's Perspective
Benjamin C. Lee
2016

A Primer on Compression in the Memory Hierarchy
Somayeh Sardashti, Angelos Arelakis, Per Stenström, and David A. Wood
2015

The Datacenter as a Computer: An Introduction to the Design of Warehouse-Scale Machines, Second edition
Luiz André Barroso, Jimmy Clidaras, and Urs Hölzle
2013

Shared-Memory Synchronization
Michael L. Scott
2013

Resilient Architecture Design for Voltage Variation
Vijay Janapa Reddi and Meeta Sharma Gupta
2013

Multithreading Architecture
Mario Nemirovsky and Dean M. Tullsen
2013

Performance Analysis and Tuning for General Purpose Graphics Processing Units (GPGPU)
Hyesoon Kim, Richard Vuduc, Sara Baghsorkhi, Jee Choi, and Wen-mei Hwu
2012

Automatic Parallelization: An Overview of Fundamental Compiler Techniques
Samuel P. Midkiff
2012

Phase Change Memory: From Devices to Systems
Moinuddin K. Qureshi, Sudhanva Gurumurthi, and Bipin Rajendran
2011

Multi-Core Cache Hierarchies
Rajeev Balasubramonian, Norman P. Jouppi, and Naveen Muralimanohar
2011

A Primer on Memory Consistency and Cache Coherence
Daniel J. Sorin, Mark D. Hill, and David A. Wood
2011

Dynamic Binary Modification: Tools, Techniques, and Applications
Kim Hazelwood
2011

Quantum Computing for Computer Architects, Second Edition
Tzvetan S. Metodi, Arvin I. Faruque, and Frederic T. Chong
2011

Quantum Computing for Computer Architects
Tzvetan S. Metodi and Frederic T. Chong
2006

Deep Learning for Computer Architects

Brandon Reagen, Robert Adolf, Paul Whatmough, Gu-Yeon Wei, and David Brooks

ISBN: 978-3-031-00628-9 paperback
ISBN: 978-3-031-01756-8 ebook

DOI 10.1007/978-3-031-01756-8

A Publication in the Springer series
SYNTHESIS LECTURES ON ADVANCES IN AUTOMOTIVE TECHNOLOGY

Lecture #41
Series Editor: Margaret Martonosi, *Princeton University*
Founding Editor Emeritus: Mark D. Hill, *University of Wisconsin, Madison*
Series ISSN
Print 1935-3235   Electronic 1935-3243

Deep Learning for Computer Architects

Brandon Reagen
Harvard University

Robert Adolf
Harvard University

Paul Whatmough
ARM Research and Harvard University

Gu-Yeon Wei
Harvard University

David Brooks
Harvard University

SYNTHESIS LECTURES ON COMPUTER ARCHITECTURE #41

ABSTRACT

Machine learning, and specifically deep learning, has been hugely disruptive in many fields of computer science. The success of deep learning techniques in solving notoriously difficult classification and regression problems has resulted in their rapid adoption in solving real-world problems. The emergence of deep learning is widely attributed to a virtuous cycle whereby fundamental advancements in training deeper models were enabled by the availability of massive datasets and high-performance computer hardware.

This text serves as a primer for computer architects in a new and rapidly evolving field. We review how machine learning has evolved since its inception in the 1960s and track the key developments leading up to the emergence of the powerful deep learning techniques that emerged in the last decade. Next we review representative workloads, including the most commonly used datasets and seminal networks across a variety of domains. In addition to discussing the workloads themselves, we also detail the most popular deep learning tools and show how aspiring practitioners can use the tools with the workloads to characterize and optimize DNNs.

The remainder of the book is dedicated to the design and optimization of hardware and architectures for machine learning. As high-performance hardware was so instrumental in the success of machine learning becoming a practical solution, this chapter recounts a variety of optimizations proposed recently to further improve future designs. Finally, we present a review of recent research published in the area as well as a taxonomy to help readers understand how various contributions fall in context.

KEYWORDS

deep learning, neural network accelerators, hardware software co-design, DNN benchmarking and characterization, hardware support for machine learning

Contents

Preface

This book is intended to be a general introduction to neural networks for those with a computer architecture, circuits, or systems background. In the introduction (Chapter 1), we define key vocabulary, recap the history and evolution of the techniques, and for make the case for additional hardware support in the field.

We then review the basics of neural networks from linear regression to perceptrons and up to today's state-of-the-art deep neural networks (Chapter 2). The scope and language is presented such that anyone should be able to follow along, and the goal is to get the community on the same page. While there has been an explosion of interest in the field, evidence suggests many terms are being conflated and that there are gaps in understandings in the area. We hope that what is presented here dispels rumors and provides common ground for nonexperts.

Following the review, we dive into tools, workloads, and characterization. For the practitioner, this may be the most useful chapter. We begin with an overview of modern neural network and machine learning software packages (namely TensorFlow, Torch, Keras, and Theano) and explain their design choices and differences to guide the reader to choosing the right tool for their work. In the second half of Chapter 3, we present a collection of commonly used, seminal workloads that we have assembled in a benchmark suite named Fathom [2]. The workloads are broken down into two categories: dataset and model, with explanation of why the workload and/or dataset is seminal as well as how it should be used. This section should also help reviewers of neural network papers better judge contributions. By having a better understanding of each of the workloads, we feel that more thoughtful interpretations of ideas and contributions are possible. Included with the benchmark is a characterization of the workloads on both a CPU and GPU.

Chapter 4 builds off of Chapter 3 and is likely of greatest interest to the architect looking to investigate accelerating neural networks with custom hardware. In this chapter, we review the Minerva accelerator design and optimization framework [114] and include details of how high-level neural network software libraries can be used in conglomeration with hardware CAD and simulation flows to codesign the algorithms and hardware. We specifically focus on the Minerva methodology and how to experiment with neural network accuracy and power, performance, and area hardware trade-offs. After reading this chapter, a graduate student should feel confident in evaluating their own accelerator/custom hardware optimizations.

In Chapter 5, we present a thorough survey of relevant hardware for neural network papers, and develop a taxonomy to help the reader understand and contrast different projects. We primarily focus on the past decade and group papers based on the level in the compute stack they address (algorithmic, software, architecture, or circuits) and by optimization type (sparsity, quantization, arithmetic approximation, and fault tolerance). The survey primarily focuses on the top machine learning, architecture, and circuit conferences; this survey attempts to capture the most relevant works for architects in the area at the time of this book's publication. The truth is there are just too many publications to possibly include them all in one place. Our hope is that the survey acts instead as a starting point; that the taxonomy provides order such that interested readers know where to look to learn more about a specific topic; and that the casual participant in hardware support for neural networks finds here a means of comparing and contrasting related work.

Finally, we conclude by dispelling any myths that hardware for deep learning research has reached its saturation point by suggesting what more remains to be done. Despite the numerous papers on the subject, we are far from done, even within supervised learning. This chapter sheds light on areas that need attention and briefly outlines other areas of machine learning. Moreover, while hardware has largely been a service industry for the machine learning community, we should really begin to think about how we can leverage modern machine learning to improve hardware design. This is a tough undertaking as it requires a true understanding of the methods rather than implementing existing designs, but if the past decade of machine learning has taught us anything, it is that these models work well. Computer architecture is among the least formal fields in computer science (being almost completely empirical and intuition based). Machine learning may have the most to offer in terms of rethinking how we design hardware, including Bayesian optimization, and shows how beneficial these techniques can be in hardware design.

Brandon Reagen, Robert Adolf, Paul Whatmough, Gu-Yeon Wei, and David Brooks
July 2017

CHAPTER 1

Introduction

Machine learning has been capturing headlines for its successes in notoriously difficult artificial intelligence problems. From the decisive victories of DeepMind's AlphaGo system against top-ranked human Go players to the wonder of self-driving cars navigating city streets, the impact of these methods are being felt far and wide. But the mathematical and computational foundations of machine learning are not magic: these are methods that have been developed gradually over the better part of a century, and it is a part of computer science and mathematics just like any other.

So what is machine learning? One way to think about it is as a way of programming with data. Instead of a human expert crafting an explicit solution to some problem, a machine learning approach is implicit: a human provides a set of rules and data, and a computer uses both to arrive at a solution automatically. This shifts the research and engineering burden from identifying specific one-off solutions to developing indirect methods that can be applied to a variety of problems. While this approach comes with a fair number its own challenges, it has the potential to solve problems for which we have no known heuristics and to be applied broadly.

The focus of this book is on a specific type of machine learning: neural networks. Neural networks can loosely be considered the computational analog of a brain. They consist of a myriad of tiny elements linked together to produce complex behaviors. Constructing a practical neural network piece by piece is beyond human capability, so, as with other machine learning approaches, we rely on indirect methods to build them. A neural network might be given pictures and taught to recognize objects or given recordings and taught to transcribe their contents. But perhaps the most interesting feature of neural networks is just how long they have been around. The harvest being reaped today was sown over the course of many decades. So to put current events into context, we begin with a historical perspective.

1.1 THE RISES AND FALLS OF NEURAL NETWORKS

Neural networks have been around since nearly the beginning of computing, but they have something of a checkered past. Early work, such as that of McCulloch and Pitts [104], was focused on creating mathematical models similar to biological neurons. Attempts at recreating brain-like behavior in hardware started in the 1950s, best exemplified by the work of Rosenblatt on his "perceptron" [117]. But interest in neural networks (both by researchers and the general public) has waxed and waned over the years. Years of optimistic enthusiasm gave way to disillusionment, which in turn was overcome again by dogged persistence. The waves of prevailing opinion are shown in Figure 1.1 overlaid on a timeline of major events that have influenced where neural networks are today. The hype generated by Rosenblatt in 1957 was quashed by Minsky and Papert in their seminal book *Perceptrons* [106]. In the book, the authors dismissed what they saw as overpromises and highlighted the technical limits of perceptrons themselves. It was famously shown that a single perceptron was incapable of learning some simple types of functions such as XOR. There were other rumblings at the time that perceptrons were not as significant as they were made out to be, mainly from the artificial intelligence community, which felt perceptrons oversimplified the difficulty of the problems the field was attempting to solve. These events precipitated the first *AI winter*, where interest and funding for machine learning (both in neural networks and in artificial intelligence more broadly) dissolved almost entirely.

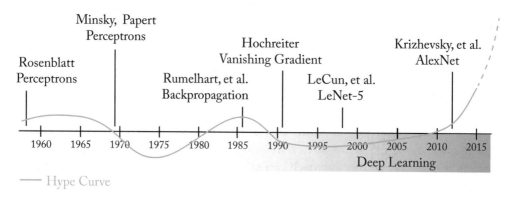

Figure 1.1: A review of the rise and fall of neural networks over years. Highlights the three major peaks: early 1960s (Rosenblatt), mid-1980s (Rumelhart), and now (deep learning).

After a decade in the cold, interest in neural networks began to pick up again as researchers started to realize that the critiques of the past may have been too harsh. A new flourishing of research introduced larger networks and new techniques for tuning them, especially on a type of computing called *parallel distributed processing*—large numbers of neurons working simultaneously to achieve some end. The defining paper of the decade came when Rumelhart, Hinton, and Williams proposed their method for backpropagation in 1986 [119]. While others have rightly been credited for inventing the technique earlier (most notably Paul Werbos who had proposed the technique twelve years before [140]), this brought backpropagation into the mainstream and changed attitudes toward neural networks. Backpropagation leveraged a simple calculus to allow for networks of arbitrary structure to be trained efficiently. Perhaps most important, this allowed for more complicated, hierarchical neural nets. This, in turn, expanded the set of problems that could be attacked and sparked interest in practical applications.

Despite this remarkable progress, overenthusiasm and hype once again contributed to looming trouble. In fact, Minsky (who partially instigated and weathered the first winter) was one of the first to warn that the second winter would come if the hype did not die down. To keep research money flowing, researchers began to promise more and more, and when they fell short of delivering on their promises, many funding agencies became disillusioned with the field as a whole. Notable examples of this include the Lighthill report from England [103] and the cancellation of the Speech Understanding Research program from DARPA in favor of more traditional systems. This downturn was accompanied by a newfound appreciation for the complexity of neural networks. As especially pointed out by Lighthill, for these models to be useful in solving real-world problems would require incredible amounts of computational power that simply was not available at the time.

While the hype died down and the money dried up, progress still moved on in the background. During the second AI winter, which lasted from the late 1980s until the mid-2000s, many substantial advances were still made. For example, the development of convolutional neural networks [89] (see Section 3.1.1) in the 1990s quietly grew out of similar models introduced earlier (e.g., the Neocognitron [48]). However, two decades would pass before widespread interest in neural networks would arise again.

1.2 THE THIRD WAVE

In the late 2000s, the second AI winter began to thaw. While many advances had been made on algorithms and theory for neural networks, what made this time around different was the setting to which neural networks awoke. As a whole, since the late 1980s, the computing landscape had changed. From the Internet and ubiquitous connectivity to smart phones and social media, the sheer volume of data being generated was overwhelming. At the same time, computing hardware continued to follow Moore's law, growing exponentially through the AI winter. The world's most powerful computer at the end of the 1980s was literally equivalent to a smart phone by 2010. Problems that used to be completely infeasible suddenly looked very realistic.

1.2.1 A VIRTUOUS CYCLE

This dramatic shift in circumstances began to drive a self-reinforcing cycle of progress and opportunity (Figure 1.2). It was the interplay between these three areas—data, algorithms, and computation—that was directly responsible for the third revival of neural networks. Each was significant on its own, but the combined benefits were more profound still.

These key factors form a virtuous cycle. As more complex, larger datasets become available, new neural network techniques are invented. These techniques typically involve larger models with mechanisms that also require more computations per model parameter. Thus, the limits of even today's most powerful, commercially available devices are being tested. As more powerful hardware is made available, models will quickly expand to consume and use every available device. The relationship between large datasets, algorithmic training advances, and high-performance hardware forms a virtuous cycle. Whenever advances are made in one, it fuels the other two to advance.

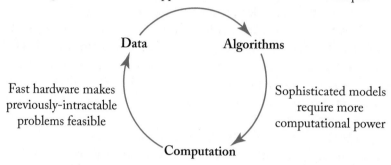

Figure 1.2: The virtuous cycle underpinning the modern deep learning revival.

Big Data   With the rise of the Internet came a deluge of data. By the early 2000s, the problem was rarely obtaining data but instead, trying to make sense of it all. Rising demand for algorithms to extract patterns from the noise neatly fit with many machine learning methods, which rely heavily on having a surfeit of data to operate. Neural networks in particular stood out: compared with simpler techniques, neural nets tend to scale better with increasing data. Of course, with more data and increasingly complicated algorithms, ever more powerful computing resources were needed.

Big Ideas   Many of the thorny issues from the late 1980s and early 1990s were still ongoing at the turn of the millennium, but progress had been made. New types of neural networks took advantage of domain-specific characteristics in areas such as image and speech processing, and new algorithms for optimizing neural nets began to whittle away at the issues that had stymied researchers in the prior decades. These ideas, collected and built over many years, were dusted off and put back to work. But instead of megabytes of data and megaflops of computational power, these techniques now had millions of times more resources to draw upon. Moreover, as these methods began to improve, they moved out of the lab and into the wider world. Success begot success, and demand increased for still more innovation.

Big Iron   Underneath it all was the relentless juggernaut of Moore's law. To be fair, computing hardware was a trend unto itself; with or without a revival of neural networks, the demand for more capability was as strong as ever. But the improvements in computing resonated with machine learning somewhat differently. As frequency scaling tapered off in the early 2000s, many application domains struggled to adjust to the new realities of parallelism. In contrast, neural networks excel on parallel hardware by their very nature. As the third wave was taking off, computer processors were shifting toward architectures that looked almost tailor-made for the new algorithms and massive datasets. As machine learning continues to grow in prevalence, the influence it has on hardware design is increasing.

This should be especially encouraging to the computer architect looking to get involved in the field of machine learning. While there has been an abundance of work in architectural support for neural networks already, the virtuous cycle suggests that there will be demand for new solutions as the field evolves. Moreover, as we point out in Chapter 3, the hardware community has only just begun to scratch the surface of what is possible with neural networks and machine learning in general.

1.3   THE ROLE OF HARDWARE IN DEEP LEARNING

Neural networks are inveterate computational hogs, and practitioners are constantly looking for new devices that might offer more capability. In fact, as a community, we are already in the middle of the second transition. The first transition, in the late 2000s, was when researchers discovered that commodity GPUs could provide significantly more throughput than desktop CPUs. Unlike many other application domains, neural networks had no trouble making the leap to a massively data-parallel programming model—many of the original algorithms from the 1980s were already formulated this way anyway. As a result, GPUs have been the dominant platform for neural networks for several years, and many of the successes in the field were the result of clusters of graphics cards churning away for weeks on end [135].

Recently, however, interest has been growing in dedicated hardware approaches. The reasoning is simple: with sufficient demand, specialized solutions can offer better performance, lower latency, lower power, or whatever else an application might need compared to a generic system. With a constant demand for more cycles from the virtuous cycle mentioned above, opportunities are growing.

1.3.1   STATE OF THE PRACTICE

With neural networks' popularity and amenability to hardware acceleration, it should come as no surprise that countless publications, prototypes, and commercial processors exist. And while it may seem overwhelming, it is in fact just the tip of the iceberg. In this section, we give a brief look at the state of the art to highlight the advances that have been made and what more needs to be done. Chapter 5 provides a more comprehensive look at the field.

To reason quantitatively about the field, we look at a commonly used research dataset called *MNIST*. MNIST is a widely studied problem used by the machine learning community as a sort of lowest common denominator. While it is no longer representative of real-world applications, it serves a useful role. MNIST is small, which makes it easier to dissect and understand, and the wealth of prior experiments on the problem means comparison is more straightforward. (An excellent testimonial as to why datasets like MNIST are imperative to the field of machine learning is given by Roger Grosse [56].)

Figure 1.3 shows the results of a literature survey on hardware and software implementations of neural networks that can run with the MNIST dataset. The details of the problem are not important, but the basic idea is that these algorithms are attempting to differentiate be-

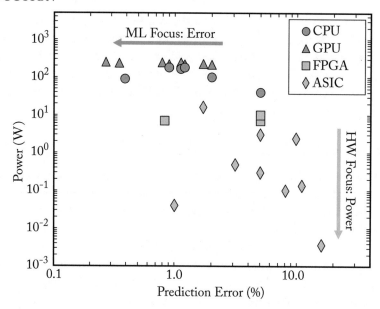

Figure 1.3: A literature survey of neural networks reveals the disconnect between machine learning research and computer architecture communities when designing neural network systems. References: CPUs [47, 63, 112, 130, 139], GPUs [24, 30, 32, 63, 128, 130, 139], FPGAs [47, 57], and ASICs [7, 18, 24, 46, 47, 83, 88, 132].

tween images of single handwritten digits (0–9). On the x-axis is the prediction error of a given network: a 1% error rate means the net correctly matched a 99% of the test images with their corresponding digit. On the y-axis is the power consumed by the neural network on the platform it was tested on. The most interesting trend here is that the machine learning community (blue) has historically focused on minimizing prediction error and favors the computational power of GPUs over CPUs, with steady progress toward the top left of the plot. In contrast, solutions from the hardware community (green) trend toward the bottom right of the figure, emphasizing practical efforts to constrain silicon area and reduce power consumption at the expense of nonnegligible reductions in prediction accuracy.

The divergent trends observed for published machine learning vs. hardware results reveal a notable gap for implementations that achieve competitive prediction accuracy with power budgets within the grasp of mobile and IoT platforms. A theme throughout this book is how to approach this problem and reason about accuracy-performance trade-offs rigorously and scientifically.

This book serves as a review of the state of the art in deep learning, in a manner that is suitable for architects. In addition, it presents workloads, metrics of interest, infrastructures, and future directions to guide readers through the field and understand how different works compare.

Figure 1.4: A photo of one of the first neural network prototypes, built by Frank Rosenblatt at Cornell. Many early neural networks were designed with specialized hardware at their core: Rosenblatt developed the Mark I Perceptron machine in the 1960s, and Minsky built SNARC in 1951. It seems fitting now that computer architecture is once again playing a large role in the lively revival of neural networks.

CHAPTER 2

Foundations of Deep Learning

2.1 NEURAL NETWORKS

The seat of human thought has long inspired scientists and philosophers, but it wasn't until the 20th century that the mechanisms behind the brain began to be unlocked. Anatomically, the brain is a massive network of simple, interconnected components: a neural network. Thought, memory, and perception are products of the interactions between the elements of this network, the result of thousands of signals being exchanged. Modern deep learning methods can trace their origins to computational analogs of biological neural networks, so it helps to have a basic understanding of their functionality.

Figure 2.1: Local neuron connectivity in a mouse brain. True color image produced by induced expression of fluorescent protein [99]. Each colored patch is a single neuron cell.

2.1.1   BIOLOGICAL NEURAL NETWORKS

The fundamental unit of the brain is a neuron (Figure 2.2a), a specialized cell that is designed to pass electrochemical signals to other parts of the nervous system, including other neurons. A neuron is comprised of three basic pieces: the soma, or body, which contains the nucleus, metabolic machinery, and other organelles common to all cells; dendrites, which act as a neuron's inputs, receiving signals from other neurons and from sensory pathways; and an axon, an elongated extension of the cell responsible for transmitting outgoing signals. Connections between neurons are formed where an axon of one neuron adjoins the dendrite of another at an intercellular boundary region called a *synapse*. The long, spindly structure of neurons enables a dense web, with some neurons having up to 200,000 different direct connections [102].

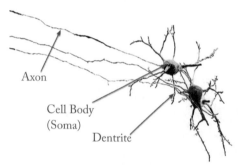

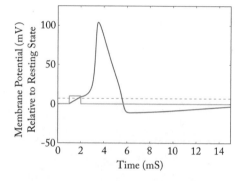

(a) Annotated visualization of the structure of a biological neuron, reconstructed from electron microscope images of 30 nm slices of a mouse brain [138].

(b) The shape of an action potential. A small external voltage stimulus (blue) triggers a cascade of charge buildup inside a neuron (red) via voltage-gated ion channels. The activation threshold is shown as a dotted line. Simulated using a Hodgkin-Huxley model of a neuron [73] (Nobel prize in Medicine, 1963).

Figure 2.2: The structure and operation of biological neurons.

Unlike most biological signaling mechanisms, communication between two neurons is an electrical phenomenon. The signal itself, called an *action potential*, is a temporary spike in local electrical charge difference across the cell membrane (see Figure 2.2b). This spike is the result of specialized membrane proteins called *voltage-gated channels*, which act as voltage-controlled ion pumps. At rest, these ion channels are closed, but when a small charge difference builds up (either through the result of a chemical stimulus from a sensory organ or from an induced charge from another neuron's synapse), one type of protein changes shape and locks into an open position, allowing sodium ions to flow into the cell, amplifying the voltage gradient. When the voltage reaches a certain threshold, the protein changes conformation again, locking itself into a closed state. Simultaneously, a second transport protein switches itself open, allowing potassium ions to flow out of the cell, reversing the voltage gradient until it returns to its resting state. The locking behavior of voltage-gated channels causes hysteresis: once a spike has started, any further external stimuli will be dominated by the rush of ions until the entire process has returned to its rest state. This phenomenon, combined with the large local charge gradient caused by the ion flow, allows an action potential to propagate unidirectionally along the surface of a neuron. This forms the fundamental mechanism of communication between neurons: a small signal received at dendrites translates into a wave of charge flowing through the cell body and down the axon, which in turn causes a buildup of charge at the synapses of other neurons to which it is connected.

The discriminative capacity of a biological neuron stems from the mixing of incoming signals. Synapses actually come in two types: excitatory and inhibitory, which cause the outgoing action potential sent from one neuron to induce either an increase or decrease of charge on the dendrite of an adjoining neuron. These small changes in charge accumulate and decay over short time scales until the overall cross-membrane voltage reaches the activation threshold for the ion channels, triggering an action potential. At a system level, these mutually induced pulses of electrical charge combine to create the storm of neural activity that we call thought.

2.1.2   ARTIFICIAL NEURAL NETWORKS

Computational and mathematical models of neurons have a long, storied history, but most efforts can be broadly grouped into two categories: (1) models that replicate biological neurons to explain or understand their behavior or (2) solve arbitrary problems using neuron-inspired models of computation. The former is typically the domain of biologists and cognitive scientists, and computational models that fall into this group are often described as *neuromorphic computing*, as the primary goal is to remain faithful to the original mechanisms. This book deals exclusively with the latter category, in which loosely bio-inspired mathematical models are brought to bear on a wide variety of unrelated, everyday problems. The two fields do share a fair amount of common ground, and both are important areas of research, but the recent rise in practical application of neural networks has been driven largely by this second area. These techniques do not solve learning problems in the same way that a human brain does, but, in exchange, they can offer other advantages, such as being simpler for a human to build or mapping more naturally to modern microprocessors. For brevity, we use "neural network" to refer to this second class of algorithms in the rest of this book.

We'll start by looking at a single artificial neuron. One of the earliest and still most widely used models is the *perceptron*, first proposed by Rosenblatt in 1957. In modern language, a perceptron is just two terms: a weighted sum of inputs x_i, and a nonlinear *activation function*, φ:

$$y = \varphi \left(\sum_i w_i x_i \right)$$

The vestiges of a biological neuron are visible: the summation reflects charge accumulation in the soma, and φ loosely models the activation threshold of voltage-gated membrane proteins. Rosenblatt, himself a psychologist, developed perceptrons as a form of neuromorphic computing, as a way of formally describing how simple elements collectively produce higher-level behavior [118]. However, individual perceptrons can just as easily be viewed as generalized linear classifiers when φ has discrete outputs: $\sum_i w_i x_i$ is a linear combination of inputs and φ is a mapping function. For instance, the linear classifier in Figure 2.3a is a weighted sum of the x- and y-coordinates of the points, and φ is just a comparison to a fixed threshold (sums above the threshold are red; those below, blue). In point of fact, many different interpretations of this equation exist, largely because it is so simple. This highlights a fundamental principle of neural networks: the ability of a neural net to model complex behavior is not due to sophisticated neurons, but to the aggregate behavior of many simple parts.

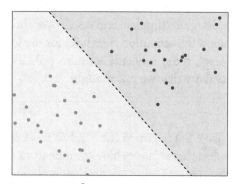

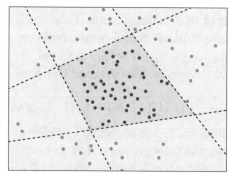

(a) Points in $\mathbb{R}^2$, subdivided by a single linear classifier. One simple way of understanding linear classifiers is as a line (or hyperplane, in higher dimensions) that splits space into two regions. In this example, points above the line are mapped to class 1 (red); those below, to class 0 (blue).

(b) Points in $\mathbb{R}^2$, subdivided by a combination of four linear classifiers. Each classifier maps *all* points to class 0 or 1, and an additional linear classifier is used to combine the four. This hierarchical model is strictly more expressive than any linear classifier by itself.

Figure 2.3: Simple elements can be combined to express more complex relationships. This is one basic tenet of deep neural networks.

The simplest neural network is called a *multilayer perceptron* or MLP. The structure is what it sounds like: we organize many parallel neurons into a layer and stack multiple layers together in sequence. Figure 2.4 shows two equivalent visual representations. The graph representation emphasizes connectivity and data dependencies. Each neuron is represented as a node, and each weight is an edge. Neurons on the left are used as inputs to neurons on their right, and values "flow" to the right. Alternately, we can focus on the values involved in the computation. Arranging the weights and input/outputs as matrices and vectors, respectively, leads us to a matrix representation. When talking about neural networks, practitioners typically consider a "layer" to encompass the weights on incoming edges as well as the activation function and its output. This can sometimes cause confusion to newcomers, as the inputs to a neural network are often referred to as the "input layer", but it is not counted toward the overall depth (number of layers). For instance, the neural network in Figure 2.4 is considered a two-layer MLP. A useful perspective is to interpret this as referring to the number of weight matrices, or to the number of layers of weighted sums (easier to see in Figure 2.4b). Similarly, since the number of layers is *depth*, we call the number of neurons in a given layer the *width* of that layer. Width and depth are often used loosely to describe and compare the overall structure of neural networks, and we return to the significance of these attributes in Section 2.1.3.

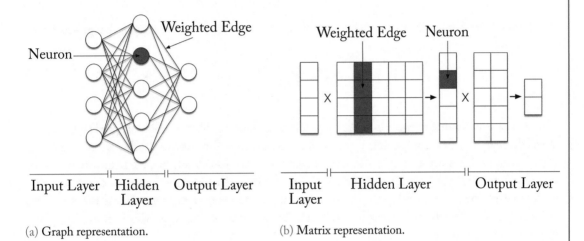

(a) Graph representation.   (b) Matrix representation.

Figure 2.4: Two ways of visualizing the same multilayer perceptron.

Visualizations can be useful mental models, but ultimately they can be reduced to the same mathematical form. Building off the expression for a single perceptron, the nth layer of an MLP can be expressed as:

$$x_{n,i} = \varphi \left(\sum_j w_{n,i,j} x_{n-1,j} \right)$$

or in vector notation,

$$\mathbf{x}_n = \varphi \left(\mathbf{w}_n \mathbf{x}_{n-1} \right)$$

We return to this expression over the course of this book, but our first use helps us understand the role and necessity of the activation function. Historically, many different functions have been chosen for φ. The very first were step-functions ⌐_, but these were superseded by continuous variants like the sigmoid ⟋ and hyperbolic tangent ⟋. Modern neural nets tend to use a simpler function called a *rectified linear unit* (ReLU) ⟋. A natural question to ask when encountering φ for the first time is Why is it even required? The math provides a straightforward, satisfying answer. Assume for a minute that we replaced φ with an identity function, how would a two-layer MLP behave? We start with the expressions for both layers, using $\varphi(x) = x$:

$$\mathbf{x}_1 = \mathbf{w}_1 \mathbf{x}_0$$
$$\mathbf{x}_2 = \mathbf{w}_2 \mathbf{x}_1$$

Now we simply substitute

$$\mathbf{x}_2 = \mathbf{w}_2 (\mathbf{w}_1 \mathbf{x}_0)$$

Since matrix multiplication is associative and $\mathbf{w}_n$ are constant, we can reduce the product to a new matrix

$$\mathbf{x}_2 = \mathbf{w}' \mathbf{x}_0$$

This leaves us with an aggregate expression, which has the same form as each original layer. This is a well-known identity: composing any number of linear transformations produces another linear transformation. It implies that without a *nonlinear* activation function φ, a neural network of any depth is identical to some single-layer network. In other words, without φ, depth is a just a pretense of complexity. What might be surprising is that even simple nonlinear functions are sufficient to enable complex behavior. The definition of the widely used ReLU function is just the positive component of its input:

$$\varphi_{\text{ReLU}}(x) = \begin{cases} x \leq 0: & 0 \\ x > 0: & x \end{cases}$$

In practice, neural networks can become much more complicated than the multilayer perceptrons introduced here, and often they contain a variety of types of layers (see Chapter 3). Fundamentally, though, the underlying premise remains the same: complexity is achieved through a combination of simpler elements broken up by nonlinearities.

2.1.3   DEEP NEURAL NETWORKS

As a linear classifier, a single perceptron is actually a fairly primitive model, and it is fairly easy to come up with noncontrived examples for which a perceptron fairs poorly, the most famous being boolean XOR [105]. Historically, this was a source of contention in the AI community in the late 1960s, and it was in part responsible for casting doubt on the viability of artificial neural networks for most of the 1970s. Thus, it was a surprise in 1989 when several independent researchers discovered that even a single hidden layer is provably sufficient for an MLP to express any continuous function with arbitrary precision [37, 49, 74]. Unfortunately, this result, called the *Universal Approximation Theorem*, says nothing about how to construct or tune such a network, nor how efficient it will be in computing it. As it turns out, expressing complicated functions (especially in high dimensions) is prohibitively expensive for a two-layer MLP—the hidden layer needs far too many neurons.

Instead, researchers turned to networks with multiple narrower layers: deep neural networks, or DNNs. Deep neural networks are attractive in principle: when solving a complicated problem, it is easier to break it into smaller pieces and then build upon those results. This approach is faster and easier for humans, but a more important advantage (from a computational point of view) is that these smaller, intermediate solutions can often be reused within the same neural network. For instances, in understanding spoken language, it makes sense to recognize common phonetic patterns instead of attempting to recognize each word directly from a waveform. In practice, modern neural networks have largely succeeded in replicating this type of behavior. In fact, in some cases, it is possible to see exactly what kind of intermediate patterns are being constructed by interrogating hidden layers and analyzing the output [81, 131, 146].

The trade-off and challenge with deep neural networks lies in how to tune them to solve a given problem. Ironically, the same nonlinearities that give an MLP its expressivity also preclude conventional techniques for building linear classifiers. Moreover, as researchers in the 1980s and 1990s discovered, DNNs come with their own set of problems (e.g., vanishing gradient problem explained in Section 2.2.2). Still, despite their complications, DNNs eventually ended up enabling a wide variety of new application areas and setting off a wave of new research, one that we are still riding today.

2.2 LEARNING

While have looked at the basic structure of deep neural networks, we haven't yet described how to convince one to do something useful like categorize images or transcribe speech. Neural networks are usually placed within the larger field of machine learning, a discipline that broadly deals with systems the behavior of which is directed by data rather than direct instructions. The word *learning* is a loaded term, as it is easily confused with human experiences such as learning a skill like painting, or learning to understand a foreign language, which are difficult even to explain—what does it actually mean to learn to paint? Moreover, machine learning algorithms are ultimately still implemented as finite computer programs executed by a computer.

So what does it mean for a machine to learn? In a traditional program, most of the application-level behavior is specified by a programmer. For instance, a finite element analysis code is written to compute the effects of forces between a multitude of small pieces, and these relationships are known and specified ahead of time by an expert sitting at a keyboard. By contrast, a machine learning algorithm largely consists of a set of rules for using and updating a set of parameters, rather than a rule about the correct values for those parameters. The linear classifier in Figure 2.3a has no rule that specifies that red points tend to have more positive coordinates than blue. Instead, it has a set of rules that describes how to use and update the parameters for a line according to labeled data provided to it. We use the word *learn* to describe the process of using these rules to adjust the parameters of a generic model such that it optimizes some objective.

It is important to understand that there is no magic occurring here. If we construct a neural network and attempt to feed it famous paintings, it will not somehow begin to emulate an impressionist. If, on the other hand, we create a neural network and then score and adjust its parameters based on its ability to simultaneously reconstruct both examples of old impressionist artwork and modern scenery, it *can* be coaxed into emulating an impressionist (Figure 2.5 [51]). This is the heart of deep learning: we are not using the occult to capture some abstract concept; we are adjusting model parameters based on quantifiable metrics.

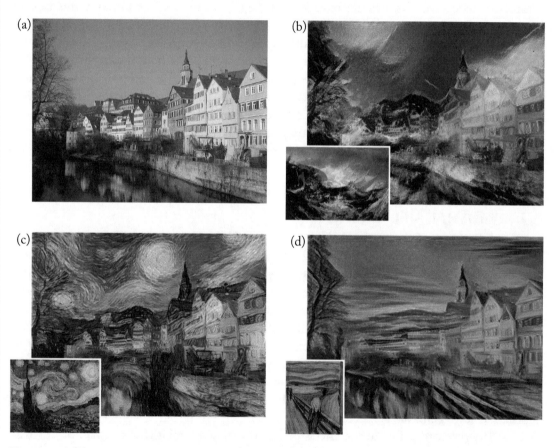

Figure 2.5: The results of a neural network constructed to combine different artistic styles with arbitrary photographs. The source image (a) is combined with different art samples to produce outputs with elements of each. Work and images courtesy of Gatys, Ecker, and Bethge [51].

2.2.1 TYPES OF LEARNING

Because machine learning models are driven by data, there are a variety of different tasks that can be attacked, depending on the data available. It is useful to distinguish these tasks because they heavily influence which algorithms and techniques we use.

The simplest learning task is the case where we have a set of matching inputs and outputs for some process or function and our goal is to predict the output of future inputs. This is called *supervised learning*. Typically, we divide supervised learning into two steps: *training*, where we tune a model's parameters using the given sample inputs, and *inference*, where we use the learned model to estimate the output of new inputs:

$$\mathbf{x}, \mathbf{y} \to M \qquad \text{Training phase}$$
$$M(x') \to y' \qquad \text{Inference phase}$$

where M represents the model that is being trained to map $\mathbf{x}$ to $\mathbf{y}$. The model can then infer an output y' from an unseen input x'. Regression and classification are examples of supervised learning problems.

Its counterpart, *unsupervised learning*, is simply the same problem except without sample outputs:

$$\mathbf{x} \to M \qquad \text{Training phase}$$
$$M(x') \to y' \qquad \text{Inference phase}$$

Because there is no given output, y can take a variety of forms. For instance, clustering is a form of unsupervised learning where the output of M is a label for every input x (its associated cluster). Outlier detection is another example, where M is a binary classifier (i.e., "is this point an outlier or not?"). *Generative models* are a related concept: these are used to produce new samples from a population defined by a set of examples. Generative models can be seen as a form of unsupervised learning where no unseen input is provided (or more accurately, the unseen input is a randomized configuration of the internal state of the model):

$$\mathbf{x} \to M \qquad \text{Training phase}$$
$$M \to y' \qquad \text{Inference phase}$$

where y' is of the same type as the elements of $\mathbf{x}$. Generative models can be used to generate entirely new data, such as the stylized paintings in Figure 2.5.

There are more complicated forms of learning as well. *Reinforcement learning* is related to supervised learning but decouples the form of the training outputs from that of the inference output. Typically, the output of a reinforcement learning model is called an *action*, and the label for each training input is called a *reward*:

$$\mathbf{x}, \mathbf{r} \to M \qquad \text{Training phase}$$
$$M(x') \to y' \qquad \text{Inference phase}$$

In reinforcement learning problems, a reward may not correspond neatly to an input—it may be the result of several inputs, an input in the past, or no input in particular. Often, reinforcement learning is used in online problems, where training occurs continuously. In these situations, training and inference phases are interleaved as inputs are processed. A model infers some output action from its input, that action produces some reward from the external system (possibly nothing), and then the initial input and subsequent reward are used to update the model. The model infers another action output, and the process repeats. Game-playing and robotic control systems are often framed as reinforcement learning problems, since there is usually no "correct" output, only consequences, which are often only loosely connected to a specific action.

2.2.2   HOW DEEP NEURAL NETWORKS LEARN

So far, we have been vague about how a model's parameters are updated in order for the model to accomplish its learning task; we claimed only that they were "based on quantifiable metrics." It is useful to start at the beginning: what does a deep neural network model look like before it has been given any data? The basic structure and characteristics like number of layers, size of layers, and activation function, are fixed—these are the rules that govern how the model operates. The values for neuron weights, by contrast, change based on the data, and at the outset, all of these weights are initialized randomly. There is a great deal of work on exactly what distribution these random values should come from [52, 65], but for our purposes, it is sufficient to say that they should be small and not identical. (We see why shortly.)

An untrained model can still be used for inference. Because the weights are selected randomly, it is highly unlikely that the model will do anything useful, but that does not prevent us from running it. Let's assume we have a set of training inputs $\mathbf{x}, \mathbf{y}$. For simplicity, we assume that we are solving a supervised learning task, but the principle can be extended to other types of learning as well. With a randomly initialized model M, we can produce an estimate $\hat{\mathbf{y}}$, which is the output of $M(\mathbf{x})$. As we noted above, this value is unlikely to look anything like the true values $\mathbf{y}$. However, since we now have a set of true outputs and our model's corresponding estimate, we can compare the two and see how far off our model was. Intuitively, this is just a fancy form of guess-and-check: we don't know what the right weights are, so we guess random values and check the discrepancy.

Loss Functions   One of the key design elements in training a neural network is what function we use to evaluate the difference between the true and estimated outputs. This expression is called a *loss function*, and its choice depends on the problem at hand. A naive guess might just be to take the difference between the two $L(\mathbf{y}, \hat{\mathbf{y}}) = \mathbf{y} - \hat{\mathbf{y}}$, but our goal here is to find a function that will be minimized when the two are equal—this function would cause $\hat{\mathbf{y}}$ to go to negative infinity, minimizing the loss function but doing little to help us. A second guess might be $L(\mathbf{y}, \hat{\mathbf{y}}) = |\mathbf{y} - \hat{\mathbf{y}}|$. This is also called the L1 norm of the difference, and it is not a terrible choice. In practice, many DNNs used for regression problems use the L2 norm ($L(\mathbf{y}, \hat{\mathbf{y}}) = (\mathbf{y} - \hat{\mathbf{y}})^2$), also known as *root mean squared error* (RMSE).

For classification problems, RMSE is less appropriate. Assume your problem has ten categories, and for every x there is exactly one correct class y. The first issue is that classes are not ordinal—just because class 0 is numerically close to class 1 doesn't mean they're any more similar than class 5 or 9. A common way around this is to encode the classes as separate elements in a vector. So the correct categories $\mathbf{y}$ would become a set of vectors each of length ten with a 1 in the yth place and zeroes elsewhere. (This is also called *one-hot* encoding). With this conversion in place, the classification problem can (mechanically, at least) be treated like a regression problem: the goal is again to make our model minimize the difference between two values, only in this case, they are actually vectors. For a single input x, our DNN would produce a vector of ten values $\hat{y}$, which we would then compare to the true vector y of 9 zeros and 1 one. Using RMSE as a loss function is problematic here: it tends to emphasize differences between the nine wrong categories at the expense of the right one. While we could try to tweak this loss function, most practitioners have settled on an alternate expression called *cross-entropy loss*:

$$L(\mathbf{y}, \hat{\mathbf{y}}) = -\frac{1}{n} \sum_i \ln \left(\frac{e^{y_i}}{\sum_j e^{\hat{y}_j}} \right)$$

While this looks rather complicated, it is based on a fairly straightforward idea. If we force the output of our neural network to be a probability distribution (i.e., real values between 0 and 1, and a sum of one), and we encode our classification values as a discrete binary distribution (where the correct class has value 1 and all other classes are 0—a one-hot encoding again), then the problem of comparing outputs can be viewed as that of comparing two probability distributions. It turns out that there are already good metrics for this difference. Cross-entropy is an information-theoretic measure of how many bits are needed to describe the true distribution $\mathbf{y}$ using an optimal encoding based on the distribution $\hat{\mathbf{y}}$. If $\hat{\mathbf{y}}$ is a perfect (identical) estimate of $\mathbf{y}$, then the cross-entropy will simply be the entropy of $\mathbf{y}$, which in our case is zero, since all the probability mass of $\mathbf{y}$ is on one value. Alternately, for readers already familiar with basic machine learning, cross-entropy can be thought of as a multiclass generalization of logistic regression. For a two-class problem, the two measures are identical. It's entirely possible to use a simpler measure like absolute difference or root-mean-squared error, but empirically, cross-entropy tends to be more effective for classification problems.

Optimization   So now that we have a guess (our model's estimate $\hat{y}$) and a check (our cross-entropy loss L), we need a way of adjusting our model to make better guesses in the future. In other words, we want a way of adjusting the model weights to minimize our loss function. For a simple linear classifier, it is possible to derive an analytical solution, but this rapidly breaks breaks down for larger, more complicated models. Instead, nearly every deep neural network relies on some form of stochastic gradient descent (SGD). SGD is a simple idea: if we visualize the loss function as a landscape, then one method to find a local minimum is simply to walk downhill until we cannot go any further (see Figure 2.6a).

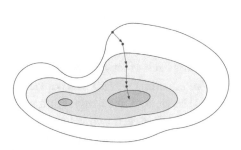

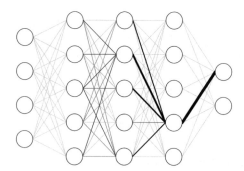

(a) Gradient descent algorithms are analogous to walking downhill to find the lowest point in a valley. If the function being optimized is differentiable, an analytical solution for the gradient can be computed. This is much faster and more accurate than numerical methods.

(b) The vanishing gradient problem. In deep neural networks, the strength of the loss gradient tends to fade as it is propagated backward due to repeated multiplication with small weight values. This makes deep networks slower to train.

Figure 2.6: In practice, deep neural networks are almost all trained using some form of stochastic gradient descent.

The gradient, however, only tells us the direction in which to shift our estimate $\hat{y}$, not how to update our neural network to realize that shift. Since we can't change $\hat{y}$ directly, we change each weight in our network so that it produces the appropriate output for us. To achieve this end, we rely on a technique called *backpropagation*. Backpropagation has a history of reinvention, but its use in training neural networks was first mentioned by Paul Werbos [140] and brought into the mainstream by a 1986 paper by Rumelhart, Hinton, and Williams [119]. The intuition behind backpropagation is that elements of a neural network should be adjusted proportionally to the degree to which they contributed to generating the original output estimate. Neurons that pushed $\hat{y}$ closer to y should be reinforced, and those that did the opposite, suppressed.

Recall that computing a gradient $\nabla L(\mathbf{y}, \hat{y})$ is just computing partial derivatives $\frac{\partial L}{\partial x_i}$ for all inputs in $\mathbf{x}$. To adjust individual weights, we simply take the component of the loss function, which corresponds to a particular weight: literally, $\frac{\partial L}{\partial w}$ for each w. Backpropagation is a mechanism for computing all of these partial loss components for every weight in a single pass.

The key enabling feature is that all deep neural networks are *fully differentiable*, meaning that every component, including the loss function, has an analytical expression for its derivative. This enables us to take advantage of the chain rule for differentiation to compute the partial derivatives incrementally, working backward from the loss function. Recall that the chain rule states that for multiple functions f, g, and h:

$$\frac{\partial}{\partial x} f(g(h(x))) = \frac{\partial f}{\partial g} \frac{\partial g}{\partial h} \frac{\partial h}{\partial x}$$

We would like to compute $\frac{\partial L}{\partial w_{i,j}}$ for every weight j in every layer i. The chain rule implies that computing the partial derivatives for layers in the front of a deep neural network requires computing the partial derivatives for latter layers, which we need to do anyway. In other words, backpropagation works by computing an overall loss value for the outputs of a network, then computing a partial derivative for each component that fed into it. These partials are in turn used to compute partials for the components preceding them, and so on. Only one sweep backward through the network is necessary, meaning that computing the updates to a deep neural network usually takes only a bit more time than running it forward.

We can use a simplified scalar example to run through the math. Assume we have a two-layer neural network with only a single neuron per layer:

$$\hat{y} = \varphi_1(w_1(\varphi_0(w_0 x)))$$

This is the same MLP formulation as before, just expanded into a single expression. Now we need to compute the loss function $L(y, \hat{y})$. In the single-variable case, cross-entropy loss just reduces to a difference, assuming that $0 \le \hat{y} \le 1$. (Typically the last-layer activation function ensures this property.) So we lay out the expression for $L(y, \hat{y})$, simplify, and substitute our MLP expression:

$$
\begin{aligned}
L(y, \hat{y}) &= -\ln \frac{e^y}{e^{\hat{y}}} \\
&= -\ln e^{y - \hat{y}} \\
&= \hat{y} - y \\
&= \varphi_1(w_1(\varphi_0(w_0 x))) - y
\end{aligned}
$$

Now we want to derive the partials with respect to w_1:

$$
\begin{aligned}
\frac{\partial L}{\partial w_1} &= \frac{\partial}{\partial w_1} (\varphi_1(w_1(\varphi_0(w_0 x))) - y) \\
&= \varphi_1'(w_1(\varphi_0(w_0 x))) \frac{\partial}{\partial w_1} w_1 \varphi_0(w_0 x) \\
&= \varphi_1'(w_1(\varphi_0(w_0 x))) \varphi_0(w_0 x)
\end{aligned}
$$

and w_0,

$$
\begin{aligned}
\frac{\partial L}{\partial w_0} &= \frac{\partial}{\partial w_0} \left(\varphi_1(w_1(\varphi_0(w_0 x))) - y \right) \\
&= \varphi_1'(w_1(\varphi_0(w_0 x))) \frac{\partial}{\partial w_0} w_1 \varphi_0(w_0 x) \\
&= \varphi_1'(w_1(\varphi_0(w_0 x))) w_1 \frac{\partial}{\partial w_0} \varphi_0(w_0 x) \\
&= \varphi_1'(w_1(\varphi_0(w_0 x))) w_1 \varphi_0'(w_0 x) \frac{\partial}{\partial w_0} w_0 x \\
&= \varphi_1'(w_1(\varphi_0(w_0 x))) w_1 \varphi_0'(w_0 x) x
\end{aligned}
$$

One key observation is to note the shared term $\varphi_1'(w_1(\varphi_0(w_0 x)))$ on each partial. This sharing extends both across *every* neuron in a layer and across every layer in deeper networks. Note also that the term $\varphi_0(w_0 x)$ is just x_1, the activity of the first layer. This means that not only is the computation of the partial derivatives at each layer reused, but that, if the intermediate results from the *forward* pass are saved, they can be reused as part of the backward pass to compute all of the gradient components.

It's also useful to look back to the claim we made earlier that weights should be initialized to nonidentical values. This requirement is a direct consequence of using backpropagation. Because gradient components are distributed evenly amongst the inputs of a neuron, a network with identically initialized weights spreads a gradient evenly across every neuron in a layer. In other words, if every weight in a layer is initialized to the same value, backpropagation provides no mechanism for those weights to deviate, effectively turning an n-node layer into a one-node layer permanently. If a neural network with linear activation functions can be said to have an illusion of depth (Section 2.1.2), then a neural network initialized to identical values has an illusion of width.

Vanishing and Exploding Gradients As backpropagation was taking hold in the 1980s, researchers noticed that while it worked very well for shallow networks of only a couple layers, deeper networks often converged excruciatingly slowly or failed to converge at all. Over the course of several years, the root cause of this behavior was traced to a property of backpropagation, summarized neatly in Sepp Hochreiter's thesis in 1991 [71], which describes what we now call the *vanishing gradient* problem (see Figure 2.6b). The fundamental issue is that as the gradient of the loss function propagates backward, it is multiplied by weight values at each layer. If these weight values are less than one, the gradient shrinks exponentially (it vanishes); if they are greater than one, it grows exponentially (it explodes).

Many solutions have been proposed over the years. Setting a hard bound on gradient values (i.e., simply clipping any gradient larger than some threshold) solves the exploding but not the vanishing problem. Some involved attempting to initialize the weights of a network such that the gradient would not diverge quickly (intuitively, attempting to balance a chaotic system) [52, 65].

The switch to using ReLU as an activation function does tend to improve things, as it allows gradients to grow above 1, unlike saturating functions like tanh or a sigmoid [53]. Different neural network architectures, like long short-term memory and residual networks (discussed in Chapter 3), also are less susceptible to the vanishing gradient problem. More modern techniques like batch normalization [76] also offer direct solutions. Ultimately, however, one of the main contributing factors was simply the introduction of faster computational resources. The vanishing gradient problem is not a hard boundary—the gradient components are still propagating; they are simply small. With Moore's law increasing the speed of training a neural network exponentially, many previously untrainable networks became feasible simply through brute force. This continues to be a major factor in advancing deep learning techniques today.

CHAPTER 3

Methods and Models

3.1 AN OVERVIEW OF ADVANCED NEURAL NETWORK METHODS

Chapter 2 gave an introduction to the most basic type of neural network, but the field contains a much wider array of techniques. While it is impossible to explore every type of network and variety of algorithm, we will cover some of the more popular methods that are currently in use. One key point to keep in mind is that these techniques do not fundamentally change what it is *possible* to learn. As we saw earlier in Section 2.1.3, neural networks are already theoretically capable of capturing any relationship. Instead, these approaches present trade-offs in order to produce a model that is more efficient in some ways: cheaper to compute, easier to train, or perhaps more robust to noise. Choosing which model or method to use for a problem is part of the art of deep learning.

In reading this section, remember that this book represents a snapshot in time. The field of deep learning is in flux, and the techniques that we introduce here are likely to be improved and eventually replaced with others in time. That said, scientific discovery is monotonic—the techniques of tomorrow will almost certainly be a product of the lessons of today.

3.1.1 MODEL ARCHITECTURES

The first area we explore is variations on the fundamental structure of a neural network. While a multilayer perceptron is straightforward to understand and use, many applications have characteristics that can be challenging for them. Changing the rules by which a network operates can be an effective way to make learning the domain easier.

Convolutional Neural Networks   In image processing and analysis tasks, one of the key hurdles is a wealth of input features: image and video data have a massive number of pixels to consider, and assigning each pixel as input to a neuron means at least as many weights to train. Doing so, however, is somewhat odd when looked at from a mathematical perspective. Essentially, each pixel is being treated as a completely independent input signal from every other pixel. In real images, this is almost never the case—nearby pixels capture pieces of the same object or location in the real world. This strong locality characteristic implies that a neural network might be made simpler. Indeed, conventional image-processing algorithms have long exploited hierarchical pipelines (e.g., SIFT [100]). In 1980, Kunihiko Fukushima proposed a variant of a neural network he called the *neocognitron* [48], which reused sets of neurons to detect various

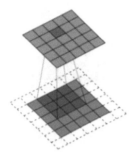

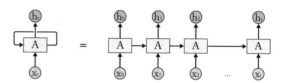

(a) Convolutional neural networks reuse small, learned filters by sliding them across an image. The computation is identical to a normal perceptron at each location.

(b) Recurrent neural networks incorporate the output of their neurons as inputs to capture sequential dependencies in input streams.

Figure 3.1: Two varieties of neural network architectures.

local features. Today, the area is dominated by *convolutional neural networks* (CNNs), a related model architecture pioneered by Yann LeCun's group at Bell labs and later NYU [89]. Intuitively, a CNN works by sliding a large number of small, trainable filters across an image (See Figure 3.1a). To train a convolutional network, the same process is applied in reverse: instead of backpropagating the loss gradient to individual neurons, we sum the gradients for every location that the same filter visited into a single value, which is then used to update the filter's weights. Mathematically, the forward operation is described by convolving a small filter with a large input image (hence the name). The backward (training) operation involves, naturally, deconvolving the gradients using the filters in order to produce a new set of gradients for the preceding layer. An alternate interpretation of a CNN is as a large MLP where many of the weights in a layer are tied together in a specific pattern. In training, when a gradient update is applied to one of these weights, it is applied to all of them. These tied (or shared) weights correspond to a convolutional filter, and the structure with which they are connected describes the shape of the filter. In this interpretation, the computation of a CNN can be thought of as a compact representation of a larger MLP with identical mathematical properties.

Modern CNNs come in several varieties as well, some of which are introduced in more depth in Section 3.2. One of the first was the five-layer "AlexNet" model, introduced in 2012 [85]. This was one of the first models to heavily rely on GPUs for training, and it brought together several novel methods. Its real impact, however, was the attention that it drew by beating a large field of human-crafted image recognition algorithms by a significant margin in a head-to-head competition [120]. In subsequent years, that same competition saw rapid developments in CNN design. Google researchers introduced *inception* networks in 2014, which substituted groups of smaller, simpler filters to reduce computational cost while maintaining expressivity [136]. A year later, a team at Microsoft Research Asia proposed *residual networks* [64], which leveraged duplicate links across convolutional layers, an idea also independently investigated by Srivastava, Greff, and Schmidthuber [129]. (Subsequent research examined the differences and relative performance of these variants [66].)

Recurrent Neural Networks   Another area that causes MLPs trouble is that of domains with input dependencies, such as speech and language. Dependencies are relationships between inputs that change their meaning. For instance, in the sentence "Emil _____ to _____", filling the first blank with "swam" implies a very restricted set of values that are likely to appear in the second blank, whereas filling it with "learned" is associated with a completely different set. This is complicated by the fact that these dependencies are contextual, not positional (replacing "Emil" with "my friend" or "a man in a hat" changes which neurons see these values). *Recurrent neural networks* (RNNs) are designed to deal with the sequential dependencies that often arise in time-series inputs. An RNN is a simple idea with complicated repercussions: neurons use outputs as inputs in some way. In the simplest case, this might mean that a single neuron adds its own weighted output to a new input, as in Figure 3.1b. As more inputs are processed, the neuron produces more outputs, which are passed back into itself again and again. RNNs that operate in this fashion are typically called *unidirectional* recurrent nets. Another variant involves several neurons that exchange their outputs across time, some forward and some backward. These *bidirectional* networks require the entire input stream to be processed before they can produce output.

Two challenges arise when using recurrent networks: first, while a recurrent neuron might be able to use output from previous inputs, it cannot easily handle long-range dependencies. For example, imagine the sentences "_____ gave me a compliment. I should thank _____." A pronoun in the second blank depends directly on the subject in the first. A unidirectional RNN would need to learn to pass on the characteristics of the subject of the phrase through many inputs. Second, because the output of an RNN depends (directly or indirectly) on many inputs, the normal backpropagation algorithm no longer works. The first challenge was the motivation behind the development of a variety of recurrent neurons, the most popular of which is the *long short-term memory* or LSTM developed by Hochreiter and Schmidhuber [72]. LSTM neurons have a separate set of connections that propagate a signal to themselves and a set of "gates" that allow new values to be set or cleared from these connections. This acts as a form of memory, and practically speaking, LSTMs are much more effective at capturing long-range dependencies. More recently, a simplified form of this neuron, the *gated recurrent unit*, has also proven popular [28].

Other Model Architectures   Various other domains have spawned more exotic neural network topologies. One popular structure is the bipyramidal encoder/decoder model, which stacks two neural networks back to back, the first that gradually reduces the size of its layer outputs and the second that gradually increases them again. The goal of this style of network is to learn a compact representation of an input—a projection—which can then be expanded into a more complex output. One direct example of this is the *autoencoder*. Autoencoders are symmetric neural nets, where the input and output for each training example are the same sample. When an autoencoder is trained, its loss function is effectively measuring how well it can recreate its input. These networks have been applied to compression (where the intermediate representation is the compressed object), but they can also be used for more general feature representation problems by exploring the semantic meaning of each projected feature.

Another recent model architecture worth mention are *read/write networks*. Like recurrent neural networks, read/write nets attempt to solve the problem of dependencies in inputs. Unlike RNNs, however, the dependencies they are designed to capture are not sequential but arbitrary. Consider the problem of sorting a sequence of numbers: while the input has some arbitrary order, the true dependencies between values depend on their shuffled positions (a 12 should end up adjacent to a 13, but their positions in the input are random). Read/write networks use an auxiliary storage structure to "remember" information for later. Memory networks [141] and neural Turing machines [54] are the two most popular examples of read/write networks.

3.1.2   SPECIALIZED LAYERS

Even within a model, modern neural networks utilize a variety of algorithmic tricks beyond perceptrons. Most of these tricks are described as layers, even though they may or may not contain anything that is trainable. Keep in mind that because deep neural networks depend on backpropagation for learning, every component that is used must be differentiable to satisfy the chain rule.

The most popular nonneuron layer is probably *pooling*. Common in image-processing tasks, pooling layers simply aggregate the outputs of neurons from one layer. The operation itself is usually very simple: a maximum, minimum, or mean of several values. Computationally, this can save an enormous amount of time and resources, since multiplicatively decimating the number of inputs to a layer persists through later stages. In a normal MLP, this is implicit, since a layer with fewer neurons than its predecessor naturally has fewer outputs. In a convolutional neural network, where pooling is most often used, this is not the case. As a single filter is convolved with an input, the number of outputs remains the same, since every input also corresponds to the center of one location at which the filter was applied. In practice, since multiple filters are generally used, the output of a convolutional layer would naturally be larger than its input. Pooling takes advantage of the same locality that convolutional filters do: if a filter records a high local response at one location, chances are that it will also have a similar response to the inputs immediately adjacent. For example, if an early-stage filter has learned to detect a striped pattern, its output will be consistently high over the entirety of a striped region. Intuitively, pooling simply represents the aggregate signal from a region, since convolutional filters are already intended to summarize local behavior.

Another common method is normalization. Normalization layers jointly transform the outputs of many neurons to conform to some constraint. Typically, this is implemented as stretching or shifting some or all of the outputs in a layer, and it is used to keep values within a similar range. For instance, Krizhevsky, Sutskever, and Hinton introduced a technique called *local response normalization*, which multiplied each value by a fractional norm of its local neighbors, which they claim improved their network's performance on unseen inputs [85]. More recently, a technique called *batch normalization* has been used to assist in training very deep neural networks [76]. Unlike most other normalization techniques, batch normalization transforms the values of a layer based on the norm across several input samples during training. This idea corrects for a phenomenon called *internal covariate shift*, which is related to a vanishing gradient problem. As a network learns, its weight values change, which in turn shifts the scale and mean of the gradients produced by the network. This can cause training to become unstable or inefficient as gradients grow or shrink. Ioffe and Szegedy, the authors behind batch normalization, proposed a differentiable layer that corrects for this shift while controlling the impact on the inference behavior of the network.

3.2 REFERENCE WORKLOADS FOR MODERN DEEP LEARNING

One of the best ways to understand how deep learning algorithms work is through experimentation. For the rest of this chapter, we introduce a set of reference models to explore their computational characteristics. The goal is not only to supply a set of modern networks that can be used in architecture research, but also to provide an intuition into the fundamental features that drive their operation.

3.2.1 CRITERIA FOR A DEEP LEARNING WORKLOAD SUITE

Choose Meaningful Models   The first question is how to select the right models to reflect the state of the art. We believe this set of workloads should have three properties: representativeness, diversity, and impact. The first is clear: our choices should reflect the best of what the deep learning community has come up with. Since there are easily dozens of models that could rightly claim this status, the need to keep the size of the set to a manageable number implies a need for diversity: each model should bring something unique to the table. Finally, the third property, impact, is the degree to which a particular technique has changed the landscape of deep learning research. Since we cannot predict the state of the field in five years, we instead try to choose methods that have imparted fundamental lessons to the work that came after—lessons that continue to be relevant to future models.

Note that we do *not* attempt to capture pre- or postprocessing steps on either the input data or the model's output. Many models depend on preprocessing for robustness and generality, but these techniques fall outside of the scope of this paper and are handled adequately by existing benchmark suites.

Faithfully Reproduce the Original Work   This leads to the question of implementation. We are lucky in that so much of the research in the deep learning community has taken place in the open. The vast majority of research papers make a concerted effort to describe their topology and hyperparameters in the name of reproducibility, and many groups have opted to place reference implementations online. This is because reproducing deep learning experiments can be challenging; small changes in the choice of tuning parameters or data preparation can lead to large differences in outcome. So in crafting implementations for our chosen models, we adopt an existing implementation, if one is available; translate one from a different source language if not; or create and validate an implementation using descriptions from the original paper. All eight Fathom workloads were rewritten to adhere to a standard model interface, expose information to our profiling tools, and remove preprocessing or extraneous logging operations.

Similar reproducibility concerns apply to data sources as well. Whenever possible, we run our workloads using the same training and test data as the original paper. In cases where this is not possible (i.e., where the data is proprietary), we choose another dataset used by similar papers in the same area. So, for instance, since we cannot train Deep Speech on Baidu's massive, private collection of recorded utterances, we substitute the widely cited TIMIT corpus [50].

Leverage a Modern Deep Learning Framework   One particularly dramatic shift in the deep learning community in the last decade has been the widespread adoption of high-level programming models. These frameworks provide two main benefits: first, they abstract the underlying hardware interface away from the programmer. High-performance code is difficult and time-consuming to write, regardless of whether it is vectorized C++ on a CPU, or CUDA on a GPU. Second, they provide libraries of useful kernels contributed by a variety of people, which act as a productivity multiplier, especially in a fast-moving environment. These frameworks have changed the development landscape, largely for the better, and it is not possible to create a realistic set of deep learning workloads without taking them into account.

Unfortunately, using a high-level framework also raises questions, primarily, to what extent does the choice of framework matter and which one should we choose? The answer to the former question is, perhaps surprisingly, not all that much. This is in large part due to a case of convergent evolution amongst the most popular libraries. Consider four of the most widely used frameworks: Torch [33], a general machine learning library written in Lua; TensorFlow [1], the dataflow-based second generation of Google's DistBelief system [41]; Theano [10], a symbolic mathematics package originally from Université de Montréal; and Caffe [78], Berkeley's deep learning library with a JSON front end and C++ back end. All four share very similar high-level structure, and the authors seem to have arrived at many of the same design decisions:

- All use a simple front-end specification language, optimized for productivity.

- All use highly tuned native back-end libraries for the actual computation. For NVidia GPUs, all four leverage the cuDNN package [26].

- Most use an application-level, "compiler-esque" optimizer (Torch does not).

- Most are declarative, domain-specific languages (only Torch is imperative).

- Most provide support for automatic differentiation (Caffe's layers are hand implemented).

- All have some notion of a fundamental building block or primitive operation: TensorFlow and Theano call them *operations*, Torch calls them *modules*, and Caffe uses *layers*.

There will undeniably be differences in the performance characteristics of a model implemented in two of these libraries. However, essential traits are retained. First, the task of writing a model in any of these frameworks consists of assembling a pipeline of primitive operations, and most of these primitives have direct analogues in other frameworks. This similarity is so strong, in fact, that automatic translators exist for taking a model in one framework and emitting one in another [40, 145], and wrapper interfaces exist for automatically generating a model in several output frameworks [29]. Second, the performance of models written in these languages are largely dominated by the primitive operations they contain, not the overhead of the framework that contains them. This means that regardless of how the neural network is constructed, the performance characteristics will depend on the number, type, and organization of these operations (we show quantitative evidence of this in Section 3.3.1). Ultimately, while an argument could be made for choosing any one of these frameworks, the decision is somewhat arbitrary because of the similarities.

3.2.2 THE FATHOM WORKLOADS

We have assembled a collection of eight archetypal deep learning workloads, which we call *Fathom*. These are not toy problems—each of these models comes from a seminal work in the deep learning community and have either set new accuracy records on competitive datasets (e.g., image classification) or pushed the limits of our understanding of deep learning (e.g., agent planning). We offer a brief description of each workload below and a summary in Table 3.1.

Sequence-to-Sequence Translation   seq2seq is a recurrent neural network for solving machine translation [135]. The technique, developed at Google in 2014, uses a multilayer pipeline of long short-term memory (LSTM) neurons to extract the meaning of a sentence and then re-emit it in another language. The core neural network is comprised of three 7-neuron layers through which word tokens flow unidirectionally. The model also leverages an attention-based model for keeping track of context in the original sentence [8]. Sequence-to-sequence translation succeeded in achieving best-of-breed accuracy, but its impact is largely derived from its

Table 3.1: The Fathom Workloads

Model Name	Style	Layers	Learning Task	
seq2seq [135]	Recurrent	7	Supervised	WMT-15 [15]
	Direct language-to-language sentence translation. State-of-the-art accuracy with a simple, language-agnostic architecture.			
133				142
speech [62]	Recurrent, Full	5	Supervised	TIMIT [50]
	Baidu's speech recognition engine. Proved purely deep-learned networks can beat hand-tuned systems.			
84				90
residual [64]	Convolutional	34	Supervised	ImageNet [42]
	Image classifier from Microsoft Research Asia. Dramatically increased the practical depth of convolutional networks. ILSVRC 2015 winner.			
125				42
alexnet [85]	Convolutional, Full	5	Supervised	ImageNet [42]
	Image classifier. Watershed for deep learning by beating hand-tuned image systems at ILSVRC 2012.			
107				9

elegance and flexibility. It is a canonical example of a recurrent *encoder-decoder* model, a technique that transforms an input into a vector in high-dimensional space, called an *embedding*.

End-to-End Memory Networks   Memory networks [141] are one of two recent efforts to decouple state from structure in a neural network. The development of memory networks stemmed from the difficulty that stateful neurons have in capturing long-range dependencies. Facebook's AI research group solved this problem by joining an indirectly addressable memory with a neural network, resulting in a model that can explicitly store and recall information. End-to-end memory networks [133] are an extension that removes the need for type annotations on inputs and dramatically streamlines training. The bAbI [142] question-answer dataset is a natural language reasoning problem, in which a model must make simple logical deductions from an unordered sequence of statements.

Deep Speech   Deep Speech was Baidu Research's attempt at a scalable speech recognition model [62]. The model is five fully connected layers of 2,048 neurons, each with one bidirectional recurrent layer. Deep Speech is a pure deep learning algorithm in that it uses spectrograms directly as inputs and learns to transcribe phonemes (as opposed to using a hand-tuned acoustic model or hidden markov model as a preprocessing stage). Its connectionist temporal classification (CTC) loss function can learn from unsegmented data, significantly reducing the cost of producing training data [55]. Deep Speech was also notable for its emphasis on efficiency: the researchers explicitly designed the model to perform well on a GPU. We implemented the Deep Speech architecture using smaller window and embedding sizes to account for differences in the TIMIT dataset [50].

Variational Autoencoder   Autoencoders are a flexible, unsupervised model often used for dimensionality reduction, feature extraction, or generating data [68]. The fundamental assumption is that there exists a compact representation of all realistic inputs (called an embedding) that can be used to both analyze and synthesize data. Variational autoencoders, invented by Kingma and Welling in 2013, make a statistical assumption about the properties of this embedding to learn to efficiently reconstruct their inputs [84]. These models are somewhat unique in that they require stochastic sampling as part of inference, not just training.

Residual Networks   Residual networks were a landmark in enabling very deep neural networks [64]. Researchers at Microsoft Research Asia in 2015 confronted the phenomenon where increasing the depth of a model degrades both training *and* validation error. Their solution was to add additional identity connections across every pair of convolutional layers, effectively training these layers on the difference between its input and output. This tactic enabled them to train models over 150 layers deep, almost seven times larger than the previous state of the art, and it won them all five tracks of the 2015 ImageNet Large-Scale Visual Recognition Competition (ILSVRC).

VGG-19   vgg is an implementation of the 19-layer convolutional network developed by the Visual Geometry Group at Oxford [125]. The success of AlexNet inspired deeper convolutional networks, and VGG was one such offspring. The key insight by Simonyan and Zisserman was that using several convolutional layers with small filters in place of a single layer with a large one makes the network easier to train. This technique both improved accuracy (winning the ILSVRC localization task and placing second in the classification task against a far more complex Google entry) and dramatically reduced the number of learnable parameters.

AlexNet   AlexNet [85] was a watershed event for the deep learning community. While now overshadowed by more advanced models, the original model made several significant contributions. Foremost, it demonstrated that an automatically trained neural network could surpass hand-tuned image classifiers by a substantial margin. It also introduced dropout as a regularization mechanism and showcased the computational power of GPUs. While a large portion of the architecture community is already working with AlexNet, its inclusion adds continuity (allowing some degree of commensurability with prior work) as well as a reference point for the other models in Fathom.

Deep Reinforcement Learning   DeepMind startled the AI community in 2013 with a deep reinforcement learning system that learned to win dozens of Atari games solely from pixels and scores, sometimes beating human experts [107, 108]. Unlike supervised algorithms, reinforcement learning improves its behavior as it receives in-game feedback, not by observing perfect play. The heart of the method is a convolutional network that selects actions using two to three convolutional layers and two to three dense layers. The model circumvented historical difficulties in extending neural networks to decoupled feedback with innovative strategies such as experience replay. We leverage the same Atari emulation environment that powered the original implementation, the Arcade Learning Environment [9].

3.3   COMPUTATIONAL INTUITION BEHIND DEEP LEARNING

These models are intended to be tools, and proper tools require skill and understanding to wield them effectively. This section is meant to build the foundation of that understanding in two ways: First, we want to provide an intuition about the behavior of deep learning workloads. It is important to understand, for instance, where time is actually spent and what the relationships are between a given model and the hardware on which it runs. Second, we want to supply a quantitative baseline on which the community can build. There is a good deal of folklore surrounding deep learning, and numbers are a good way to begin to dispel some of that.

3.3.1 MEASUREMENT AND ANALYSIS IN A DEEP LEARNING FRAMEWORK

Working with a high-level framework like TensorFlow is a double-edged sword. On the one hand, it is a complex, dynamically compiled, dataflow-oriented runtime system, so it causes problems for conventional analysis tools. Profiling at the scripting-language level with a tool like cPython is difficult because TensorFlow is declarative, so all of the actual computation is deferred. Low-level profiling (including call-path profiling) or hardware counter instrumentation with a tool like Linux's perf can provide detailed performance information, but it loses all connection to the original model: a call to a low-level library routine cannot easily be assigned to a particular layer, for instance, because those layers only exist as internal data structures. On the other hand, TensorFlow itself makes an attractive platform for measurement. The primitive operations that are used to construct a model are ideal targets for instrumentation, and a tool built around the framework already has easy access to model information like layers and parameters, so ascribing runtime behavior to model features is straightforward. Additionally, TensorFlow (like many other popular deep learning frameworks) comes with some degree of built-in tracing support. We leverage all of these features to build a set of characterization tools that focus on capturing performance information at the model level, and we use them for all of the analyses described in this chapter.

Because we use operations as the primary abstraction for understanding the performance of the Fathom models, it is worth spending time to explain them more thoroughly. (Although we only describe TensorFlow here, the principles are also applicable to the other three popular libraries mentioned in Section 4.2 and to most production-quality deep learning frameworks in general.) An operation is a node in the coarse-grained dataflow graph that defines a TensorFlow model. It is implemented as a Python function that instructs the framework to build that node, as well as a C++ function that either performs the calculation or calls down to a low-level library to do so (either the Eigen linear algebra package on a CPU or the cuBLAS or cuDNN libraries on a GPU). Operations are the smallest schedulable unit in the TensorFlow runtime, and they double as the mechanism behind its symbolic auto-differentiation support. Examples include functions like 2D matrix-matrix multiplication (MatMul), elementwise tensor exponentiation (Pow), sampling from a normal distribution (StandardRandomNormal), or computing loss function (CrossEntropy).

Decomposing models into their component operations is convenient from a performance measurement standpoint. First, operations tend to have stable, repeatable behavior across the life of a program. Most deep learning models use some variant of gradient descent and backpropagation for optimization, so programs are naturally separable on update-step boundaries (also called *minibatches* for many training problems) or between inferences. Sampling the execution time of operations across many steps allows us to quantify stability, and Figure 3.2 shows that this distribution is stationary and has low variance. Second, most deep learning models are dominated by the time spent inside their operations. Our measurements reveal that inter-operation overhead is minimal in TensorFlow: typically less than 1% to 2% of the total runtime is spent outside of operations in our workloads.

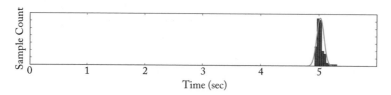

Figure 3.2: Sampling operations across the life of a program shows that their execution time is stationary and has low variance.

Finally, our experiments are carried out on a 4G Hz Skylake i7-6700 k with 32 GB RAM or an NVidia GeForce GTX 960, running TensorFlow v0.8, CUDA 7.0, and cuDNN 6.5-v2. While GPUs are more popular for the performance they provide, many frameworks (Tensor-Flow included) have incomplete support for all operations, and the fallback behavior is to run unsupported operations on the CPU, splitting execution across the PCI bus. This causes crippling performance problems, so, to avoid analysis artifacts, we opt for running most experiments on a CPU.

3.3.2 OPERATION TYPE PROFILING

The most basic performance question to ask about a workload is simply where the time is spent. Many architects working on deep learning hardware already have some sense of this, but their viewpoints are often conflicting. Some claim convolution, matrix-matrix multiplication, or matrix-vector multiplication are the predominant kernel for deep learning. To an extent, they are all right, but the truth is somewhat more nuanced. It depends on the model, environment, and use case being considered.

The general intuition about a handful of operation types dominating the overall computational time is true. While it is an exaggeration to say that a workload can be reduced to a single operation, the distribution is quite skewed, as shown in Figure 3.3. Each point on each curve represents the cumulative contribution to execution time from a single operation type. It is clear that a handful of "heavy" operation types (usually 5 to 15) are collectively responsible for upward of 90% of the programs' duration. It is important to note, however, that these types are not the same for every model (i.e., the leftmost point on each curve may represent a different operation), nor are they present in the same ratios or even at all. Figure 3.4 shows a more detailed view of the time each model spends in a given operation type. (For the sake of clarity, we only include operations with more than 1% execution time, so a given row will sum to a value somewhere between 90% and 100%.)

Unsurprisingly, convolutional neural networks are indeed dominated by convolution, and fully connected networks depend heavily on matrix multiplication. On the other hand, the break-

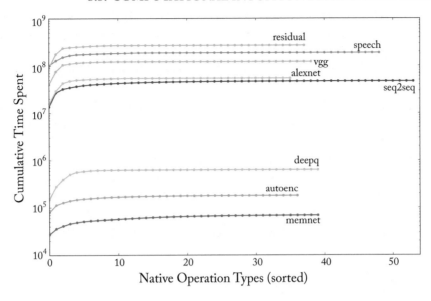

Figure 3.3: Total execution time is dominated by only a handful of unique operations.

	Add	Sub	Mul	Div	Pow	Softmax	MetMul	CrossEntropy	MaxPoolGrad	Sum	Conv2D	Conv2DBackFilter	Conv2DBackInput	RandomNormal	ApplyAdam	ApplyRMSProp	Transpose	Tile	Select	Pad	Reshape	Shape
seq2seq	3	2	35	0	0	0	32	0	0	2	0	0	0	0	0	0	3	20	0	1	0	0
memnet	2	1	33	1	0	4	12	2	0	9	0	0	0	0	5	0	9	13	0	0	1	1
speech	0	0	0	0	0	0	89	0	0	0	0	0	0	0	0	0	0	7	3	0	0	0
autoenc	3	0	6	0	5	0	58	0	0	2	0	0	0	5	8	0	0	9	0	0	0	0
residual	0	0	0	0	0	0	0	0	0	0	33	34	32	0	0	0	0	0	0	0	0	0
vgg	0	0	0	0	0	0	0	0	0	0	35	31	30	0	2	0	0	0	0	0	0	0
alexnet	0	0	0	0	0	0	7	0	3	0	31	26	31	0	0	0	0	0	0	0	0	0
deepq	0	0	0	0	0	0	11	0	0	0	33	27	20	0	0	7	0	0	0	0	0	0
	A						B	C			D			E	F		G					

Group	Op Class
A	Elementwise Arthimetic
B	Matrix Operations
C	Reduction and Expansion
D	Convolution
E	Random Sampling
F	Optimization
G	Data Movement

Figure 3.4: Breakdown of execution time by operation type for each Fathom workload.

down reveals a number of less–well-understood trends. For instance, while it is usually known that convolutional networks have gotten deeper and more expensive in recent years, it is usually not known that this has gone hand in hand with the gradual elimination of fully connected layers. Part of the justification for including alexnet, vgg, and residual is for exactly this kind of longitudinal comparison: as the winners of the ILSVRC for 2012, 2014, and 2015 respectively, they share a dataset and machine learning task, and their structures are very similar. However, `alexnet`'s two layers of locally connected neurons constitute 11% of its runtime, while `vgg`'s three fully connected layers consume only 7%, and `residual`'s single fully connected classification layer contributes less than 1%.

We can also see the effects of intentional design trade-offs. In Hannun et al.'s paper [62] describing Deep Speech, the authors describe their decision to eschew more complicated components in favor of a structurally simple, easy-to-optimize network: *"The complete RNN model...is considerably simpler than related models from the literature—we have limited ourselves to a single recurrent layer...and we do not use Long-Short-Term-Memory (LSTM) circuits.... By using a homogeneous model, we have made the computation of the recurrent activations as efficient as possible."* The evidence bears out their aims: `speech` is comprised almost exclusively of matrix-matrix multiplication operations, and the only other significant computations are part of the CTC loss function they require.

3.3.3   PERFORMANCE SIMILARITY

Operation type profiling also offers a means of assessing similarity between workloads. The mechanism is straightforward: each profile (a single row in Figure 3.4) is interpreted as a vector in high-dimensional space. Pairwise similarity can be computed using cosine similarity, and we use the inverse form $(1 - \frac{\mathbf{A} \cdot \mathbf{B}}{|\mathbf{A}||\mathbf{B}|})$ as a distance metric. We can then use agglomerative clustering with centroidal linkage to understand their relationships—i.e., we greedily group the closest two vectors, compute their centroid, and repeat until we have a single, global cluster. Figure 3.5 presents a visual representation of the hierarchical clustering generated by this process. The x-axis location of a linkage between two workloads (or two clusters of workloads) should be interpreted as a direct measure of the cosine distance between them, so the cosine distance between the centroid of the cluster containing `seq2seq` and `memnet` and that of `speech` and `autoenc` is about 0.4. High-dimensional distance is not an intuitive measure, so the value "0.4" is difficult to conceptualize, but relative distances can be understood in the normal sense: `speech` and `autoenc` share more similar performance profiles than do `seq2seq` and `memnet`.

To a deep learning expert, this dendrogram should be fairly unsurprising. The three ImageNet challenge networks are grouped closely, and `deepq`, which relies heavily on convolutional layers, is not far off. Somewhat less intuitive is the large distance between the two recurrent networks, `speech` and `seq2seq`. This is not an artifact: it is a consequence of actual underlying dissimilarity. While both networks are recurrent, Deep Speech uses CTC loss and a stack of fully connected neurons, in contrast to the stateful LSTM neurons and standard cross-entropy loss used by the sequence-to-sequence translation model. The elementwise multiplications in `seq2seq` are a result of the LSTM neurons, and the data movement operations are part of the attention-based encoder/decoder it uses [8].

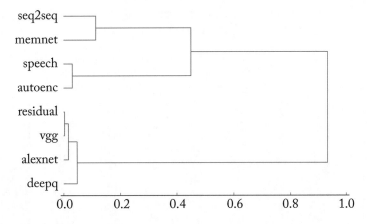

Figure 3.5: Hierarchical similarity in the Fathom workloads. The tightly clustered lower group includes all the convolutional networks.

3.3.4   TRAINING AND INFERENCE

The vast majority of deep learning systems use some variant of gradient descent and backprop-agation, which can be seen as two distinct phases: in the forward phase, the model's functions are used to compute an output value from a given input. The model's parameters are fixed. In the backward or update phase, the system evaluates its output on some criteria (a loss function). Then, for every learnable parameter in the model, the system computes the partial derivative of the loss function with respect to that parameter. This gradient is used to adjust the model pa-rameters to minimize the value of the loss function. Training a model requires evaluating both the forward and backward phases, while inference requires only the latter.

Architects working on deep learning systems generally understand that a rough symmetry exists between these two phases: most functions evaluated in the forward phase have an analogue in the backward phase with similar performance characteristics. There are exceptions to this, however, such as the evaluation of the loss function, which is only computed during the backward phase. It is still a symmetric function (e.g., the softmax function and cross-entropy loss are duals), but both parts are evaluated only during training. While this is not a revelation, it is easy to forget the performance implications of this fact. Simple classifiers tend to have fast, cheap loss functions, but many deep learning models do not, and this can cause a skew in both the overall performance of training with respect to inference and the relative importance of certain operation types.

We show the normalized training and inference runtimes for all of the Fathom workloads in Figure 3.6. Naturally, training time is more expensive than inference across the board, but the salient feature here is that it is variably faster. Convolutional networks tend to pay a slightly higher cost for training because the convolutional partial gradient involves two reduction operations in the backward phase (one for updating the filter weights and one for aggregating partials on the activity values) and only one in the forward phase.

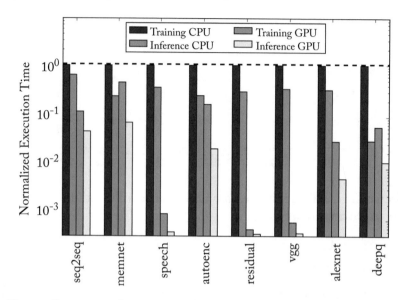

Figure 3.6: The performance of training and inference relative to the training time of each Fathom on a CPU (i.e., the lowest performance configuration). Relative performance speedups indicate the benefits each workload has whether running on a CPU or GPU.

We also evaluate both training and inference on a GPU for comparison.[1] As expected, GPU performance is substantially higher, especially on workloads with higher skew in their operation profile. GPUs also experience variability in the train-to-inference ratio across workloads, and that variability tends to be strongly correlated with that of a CPU. That is, a large gap between training and inference times on the CPU implies a similar gap on the GPU. The differences in absolute performance benefits can be partially attributed to the parallel scalability of the operations involved, which we examine in the next section.

3.3.5   PARALLELISM AND OPERATION BALANCE

Many of the most dominant operations in deep learning workloads are amenable to parallelization, and nearly all of the existing work on architectural support for learning models involves parallel hardware to some extent. The heaviest operations—convolution and matrix multiplication—are backed by an enormous body of prior work, and it is easy to overlook the effects of other operations. The breakdown in Figure 3.4 revealed that while the distribution of

[1]Running these experiments in aggregate avoids some of the detrimental effects of TensorFlow's CPU-only operations, as mentioned in Section 3.3.1. They are still present, but the effect here is just an underestimate of the total speedup over a CPU, not erroneous conclusions.

operation types is skewed, the time spent in smaller operations is not zero. As parallel resources are applied to the network, the operation types that scale strongly begin to diminish in relative importance, in accordance with Amdahl's law. As a result, new operation types begin to be more important and the profile begins to flatten out. We highlight three examples of this in Figure 3.7.

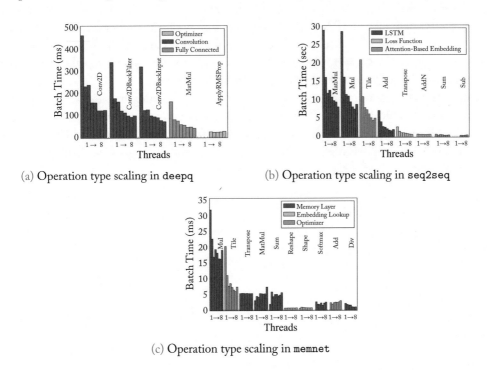

(a) Operation type scaling in `deepq`

(b) Operation type scaling in `seq2seq`

(c) Operation type scaling in `memnet`

Figure 3.7: The effect of Amdahl's law at the application level: the benefits of parallelizing matrix multiplication and convolution are limited by smaller, data-dependent operations in other parts of some models.

Each of these plots shows the absolute time spent in each operation type as we increase the amount of parallelism available within an operation (using hooks in TensorFlow to specify the available thread pool for the underlying Eigen library). We also annotate the operation types with their functional purpose in the deep learning model. For instance, in Figure 3.7a, the fourth cluster shows the time spent by `deepq` in matrix multiplication, which is the underlying operation behind the fourth and fifth fully connected layers in that model. In the extreme case of zero parallelism (single-threaded execution), the time spent in the convolutional and fully connected parts of the model dominate the execution time to the point where no other operations even show up in the profile. As parallelism increases to 8 threads, however, the optimizer (which has a large number of data-dependent operations) rises to around 7% of the execution time (which is also seen in the heatmap of Figure 3.4). A similar effect is found in the `seq2seq` model: while the LSTM neurons and attention embedding portions of the model consume an enormous amount of time at low levels of parallelism, the loss function begins to become visible at higher thread counts.

The end-to-end memory model in Figure 3.7c is a more complicated story. Many of the operations in the memory layers operate on small, skinny tensors. While these operations are frequent, they do not parallelize well (as the trip count is too low for thread-level parallelism, so the underlying library avoids it). The elementwise multiplication is an exception (it operates on the final outputs of the memory layer, which is a wide tensor), as is the optimizer, which updates learning rates for a large collection of parameters simulatenously.

The lesson here is that the performance behavior of deep learning models is inextricably tied to their application-level structure. While convolution and matrix multiplication are attractive targets for hardware support, there are limits to the benefits that can be extracted from them. This is especially true for deep learning models with nonconvolutional layers, sophisticated loss functions or optimization algorithms, or sparse storage.

C H A P T E R   4

Neural Network Accelerator Optimization: A Case Study

There are many ways to go about building custom hardware to accelerate neural networks, and many types of networks to accelerate. In this chapter, we focus on designing and optimizing hardware accelerators for executing inferences on fully connected neural networks. This narrow focus allows us to provide a detailed case study. None of the methods (or the methodology) are specific to fully connected neurons so can be applied to all other commonly used types with varying efficacy. Moreover, while this chapter specifically concerns building hardware to speed up inferences, other methods are available to accelerate training. The computational regularity of inferences makes it relatively easy to design an accelerator that performs well. What makes neural network hardware design difficult is the codesign between the hardware and neural networks' unique properties. This codesign stems from neural networks' remarkable susceptibility to approximations of their execution meaning that rather aggressive approximations can be made without sacrificing the model's predictive capabilities. To properly study this codesign, a design flow must be constructed to accurately account for approximation both at the model prediction and hardware level.

In this chapter, we present a methodology for studying the codesign and optimization of neural network accelerators. Specifically, we focus on *Minerva* [114] and walk through its construction and intended use as a case study. Minerva is a highly automated codesign flow that combines insights and techniques across the algorithm, architecture, and circuit layers, enabling low-power hardware acceleration for executing highly accurate neural networks. Minerva addresses the needs of the principled approach by operating across three layers: the algorithmic, architectural, and circuit. The Minerva flow, outlined in Figure 4.2, first establishes a fair baseline design by extensively exploring the neural network training and accelerator microarchitectural design spaces, identifying an ideal neural network topology, set of weights, and accelerator implementation. The resulting design is competitive with how neural network ASIC accelerators are designed today. Minerva then applies three cross-layer optimization steps to this baseline design, including (i) fine-grain, heterogeneous *data-type quantization*, (ii) dynamic *operation pruning*, and (iii) algorithm-aware *fault mitigation* for low-voltage SRAM operation.

The enumerated optimizations presented in the case study are used to explain the process and are implemented in Minerva today. However, Minerva is not limited to these three and is easily extensible to include other optimizations such as weight compression [61] or binary networks [34]. Minerva's multilevel approach enables researchers to quantify the model accuracy and power, performance, and area hardware trade-offs. Before presenting Minerva in detail, we first discuss the accuracy trade-offs in neural network accelerators and the importance of bounding approximations, and we propose a rigorous way to do so.

4.1   NEURAL NETWORKS AND THE SIMPLICITY WALL

Neural networks, for the most part, have all the properties of a workload extremely amenable to hardware acceleration. Consider the two prototypical neuron types: fully connected and convolutional. In both styles, the loop trip-counts are statically defined and govern all control flow and memory access patterns. This means that very few resources are needed for control and little to no speculation is required to achieve high performance. Moreover, knowing all memory accesses statically enables laying out memory optimally such that maximal parallelism can be extracted. These properties make neural networks ideal workloads for hardware acceleration. The potential for acceleration in addition to the workload's popularity has spurred researchers and engineers to quantify the performance improvements and energy savings of building such devices.

While the advances have been fantastic, further progress is inhibited by the simplicity of the workload. If we look at a breakdown of where the cycles of these models are being spent (see Chapter 3), we find that almost all the time is spent executing matrix-multiplication and convolutions—well-understood operations commonly considered for acceleration. There is still more research required to understand how to best fine-tune architectures to manage data movement and memory systems. This will be especially true as more exotic neuron styles mature.

We call this phenomenon the *Simplicity Wall*, and it is reached once all safe optimizations have been applied. A *safe optimization* is one that can be applied without restraint to improve hardware efficiency without compromising the model's, or more generally the workload's, accuracy. Examples of this would be optimizing the memory hierarchy or multiplier implementation as neither affect the result of the computation.

One way to overcome the Simplicity Wall is to approximate computation. In general, algorithmic approximations can result in substantial energy savings, e.g., [121]. Neural networks in particular provide unprecedented potential for approximation, so much so that every year dozens of publications cover the topic alone. We call all optimizations that alter the definition of the neural network model *unsafe* and further argue that to get beyond the Simplicity Wall, unsafe optimizations are necessary.

4.1.1   BEYOND THE WALL: BOUNDING UNSAFE OPTIMIZATIONS

It is important to note the difference between unsafe optimizations and approximate computing. All unsafe optimizations could be used as approximate computing techniques. However, a distinction is made as to whether or not the optimization has a *notable* impact on the output. In general, this is a very difficult metric to quantify as "notable" is open to interpretation. With respect to neural networks, we propose a general, quantitative measure. Any approximation ex-

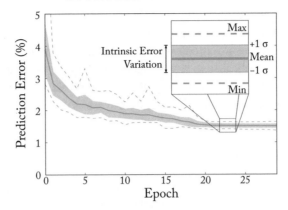

Figure 4.1: By training the same network using many random initial conditions, we can measure the intrinsic error variation in its converged state. All our optimizations are designed to have an accuracy degradation below this threshold, so their effects are indistinguishable from noise. This is the network used for the case study, which assumes an ITN of 1 error bound.

ceeding the metric is said to be approximate as the hardware improvements come at the expense of model accuracy; conversely, as long as the optimization's impact is below the threshold, it is still considered unsafe, but not approximate.

We call this metric *Iso-Training Noise* (ITN), and it is a computable score that quantifies how well an approximated model performs relative to the algorithmically correct one. It works by measuring the training noise inherent in all neural networks. The measurement is taken by training the exact same network (i.e., a network with fixed hyperparameters) multiple times using a different initial set of weight values (derived by using different initial random seeds). An example of this is shown in Figure 4.1. Once each uniquely seeded network has converged, a distribution of test error can be calculated as well as the standard deviations away from the mean across all trained models. The ITN measures implicit noise in a particular model by measuring how noisy it is in its converged state. The noise is quantified by computing standard deviations from the mean, and these are used as bounds for unsafe optimizations. ITN standard deviations are appropriate as the metric can be interpreted in this way: as long as unsafe optimizations do not exceed the bound, the unsafe optimization affects the model no more than using a different initial state.

The reason these distinctions and metrics are so important is that model accuracy is the entire reason that neural networks are relevant today. Were it not for their ability to solve difficult classification and regression problems, they would be irrelevant. This is the motivation for developing the metric and harping on the distinction between unsafe optimizations and approximate computing—if computational efficiency is gained at the expense of model fidelity, the optimization is too aggressive and the result of the computation is likely useless. However, because these are stochastic methods and there is noise, we believe working within an ITN of 1 is applicable (i.e., the optimizations impact model error no more than 1 standard deviation from the mean). There may be situations where aggressive unsafe optimizations are tolerable; in such cases, ITN is still applicable as researchers can report how far off an optimized model is from the mean. Irrespective of the use case, all unsafe optimizations demand better methods for evaluation. ITN provides a method for neural networks.

4.2   MINERVA: A THREE-PRONGED APPROACH

Research on designing and optimizing neural network accelerators requires multiple tools. To consider unsafe optimizations as discussed above, methods for working at the algorithmic to circuit layers are needed. For example, consider studying the utility of using fixed-point rather than floating-point data types, a commonly used unsafe optimization found in nearly all inference neural network accelerators today. First, we require a way to train models and manipulate them to understand the impact various fixed-point types have on accuracy.[1] Second, we need a means of quickly studying the power, performance, and area trade-offs of the various design options. Third, we need a way to characterize hardware components such that they can be used in higher level models (e.g., multiplier and SRAM costs of different fixed-point precisions). By properly combining these three tools, one for each need, researchers can rapidly explore design decisions to optimize for both high performance and model accuracy together.

The proposed design methodology works across three layers of the compute stack. At the highest level, a software model trains neural networks and emulates the impact of the optimizations on the model. Underneath the software level is an architectural simulation level. This level is used to quickly estimate power, performance, and area numbers for the hardware designs being studied in the software level. Architectural simulators take considerably longer to run than the software emulation frameworks but provide a relatively quick answer to how well the design is performing. Finally, the circuit and CAD layer is used to characterize and build libraries of hardware to be used at higher levels of abstraction. In addition, this layer can be used to validate simulation results or prototype the design in actual hardware (FPGA or ASIC).

Minerva is an implementation of this approach. Minerva uses the Keras machine learning library to train and manipulate neural networks, the Aladdin accelerator simulator to rapidly generate hardware estimates, and Design Compiler and PrimeTime to characterize model libraries and validate simulation results. Operating across three layers, Minerva is a five-stage optimization pipeline. The first two stages establish a baseline accelerator and only consider safe optimizations, and the last three consist of unsafe optimizations to reduce power and area.

---

[1]We note that while floating-point precision is in fact an approximation of true numbers, we are assuming it is algorithmically correct as it is the baseline ubiquitously assumed in neural network research.

Minerva and this chapter specifically focus on building inference accelerators. Paying attention to and discussing training help us find a proper baseline model for which to build an accelerator. While many off-the-shelf models exist, we believe that, looking forward, the optimal solutions will come from codesigning neural network hyperparameters with the microarchitectures on which they execute.

Software Level: Keras   To understand how each Minerva stage affects prediction accuracy, Minerva uses a software model built on top of the Keras [29] machine learning library. This GPU-accelerated code enables us to explore the large hyperparameter space in Stage 1. To evaluate the optimizations, we augment Keras with a software model for each technique. Evaluating fixed-point types in Stage 3 was done by building a fixed-point arithmetic emulation library and wrapping native types with quantization calls. To prune insignificant activations (Stage 4), a thresholding operation is added to the activation function of each neural network layer. This function checks each activity value and zeros all activations below the threshold, removing them from the prediction computation. To study faults in neural network weights (Stage 5), we built a fault injection framework around Keras. Before making predictions, the framework uses a fault distribution, derived from SPICE simulations for low-voltage SRAMs, to randomly mutate model weights. If a user wishes to add an unsafe optimization, it needs to be implemented in the software model.

Architecture Level: Aladdin   Stage 2 of Minerva automates a large design space exploration of microarchitectural parameters used to design an inference accelerator for the network found in stage 1 (e.g., loop level parallelism, memory bandwidth, clock frequency, etc.) in order to settle on a Pareto-optimal starting point for further optimizations. The accelerator design space that Minerva can explore is vast, easily exceeding several thousand points. To exhaustively explore this space, we rely on Aladdin, a cycle-accurate, power, performance, and area (PPA) accelerator simulator [123].

To model the optimizations in Stages 3 to 5, Aladdin is extended in the following ways: First, fixed-point data types are modeled by adding support for variable types wherein each variable and array in the C-code is tagged to specify its precision. Second, Aladdin interprets these annotations by mapping them to characterized PPA libraries to apply appropriate costs. Overheads such as additional comparators associated with operation pruning and SRAM fault detection were modeled by inserting code that is representative of the overhead into the input C-code. The benefits of operation pruning are informed by the Keras software model that tracks each elided neuronal MAC operation. This information is relayed to Aladdin and used during an added activity trace postprocessing stage where each skipped operation is removed to model dynamic power savings. To model power with respect to reduced SRAM voltages, Aladdin's nominal SRAM libraries are replaced with the corresponding low-voltage libraries.

Circuit Level: EDA To achieve accurate results, Aladdin requires detailed PPA hardware characterization to be fed into its models as represented by the arrows from the circuit to architecture levels in Figure 4.2. Aladdin PPA characterization libraries are built using PrimePower for all datapath elements needed to simulate neural network accelerators with commercial standard cell libraries in 40 nm CMOS. For SRAM modeling, we use SPICE simulations with foundry-supplied memory compilers. Included in the PPA characterization are the fixed-point types (for Stage 3) and reduced voltage SRAMs (for Stage 5). Fault distributions corresponding to each SRAM voltage reduction step are modeled using Monte Carlo SPICE simulation with 10,000 samples, similar to the methodology of [3].

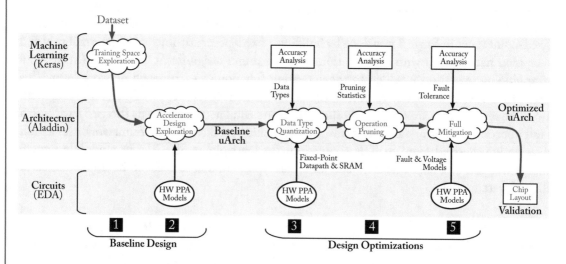

Figure 4.2: The five stages of Minerva. Analysis details for each stage and the tool-chain are presented in Section 4.2.

We validate PPA estimates for Minerva's optimized design (Stage 5 output) by comparing this design to a fully place-and-routed implementation (using Cadence SoC Encounter) of hand-written RTL informed by the parameters determined by Aladdin. The power dissipation of the accelerator is dominated by datapath and memory elements, which Aladdin models accurately using the detailed PPA libraries. Validation results from [114] show that Aladdin estimates are within 12% of a fully place-and-routed design.

Note that the simulation error in Aladdin is completely orthogonal to the neural network's prediction error. Network accuracy is computed entirely by the software layer; Aladdin is completely unaware of its existence. Aladdin models the PPA of the workload, and the error here at the architectural level is between simulation results and real hardware.

4.3   ESTABLISHING A BASELINE: SAFE OPTIMIZATIONS

In this section, we present the first two stages of Minerva.

4.3.1   TRAINING SPACE EXPLORATION

To begin, we must settle on a network and train a set of weights. There are two ways of doing this: One option is to take a network from a previous work and either download the weights directly or retrain them based on a given set of hyperparameters. The other option is to search the training space. The advantage of training our own model is that we have the power to select a network that fits our exact needs; as many network topologies are also capable of achieving near state-of-the-art accuracy, this also ensures we are starting from a fair baseline—otherwise the optimizations could be starting with an artificially large network, and the optimization gains could be inflated as they are mostly whittling down artificial network bloat.

While a training space search becomes intractable for large problems (which require large networks and take a long time to train), we believe that moving forward, bringing training into the hardware design loop is necessary. To be able to train for certain network properties intentionally—rather than simply working with fixed, unaware models—promises to offer more optimization opportunities. Also, as the optimizations and tuning become more complex, it becomes increasingly difficult for humans to reason about parameter settings and inter-parameter relationships. As an example, consider the challenge of tuning various regularization factors, the number of neurons per layer, and datatypes to both maximize the model's accuracy and minimize its energy consumption. Other machine learning techniques are looking to solve this problem; these methods are beyond the scope of this book. See recent examples in [67, 115].

Hyperparameter Space Exploration   The first stage of Minerva explores the neural network training space, identifying hyperparameters that provide optimal predictive capabilities. If the user chooses not to explore the training space, this process can be skipped by simply specifying a particular network and set of weights. To explore the hyperparameter space, Minerva considers the number of hidden layers, number of nodes per layer, and L1/L2 weight regularization penalties. The parameters considered are limited, not by Minerva but by what the software framework supports and what the user is interested in sweeping. For example, it is trivial to add hyperparameters like Dropout rate [69] and vary the activation function; this would give a richer space but require more training cycles.

Minerva then trains a neural network for each point and selects the one with the lowest prediction error and number of parameters. The weights for the trained network are then fixed and used for all subsequent experiments. Figure 4.3 plots the resulting prediction errors as a function of the number of network weights for MNIST. Larger networks often have smaller predictive error. However, beyond a certain point, the resources required to store the weights dominate the marginal increase in prediction error.

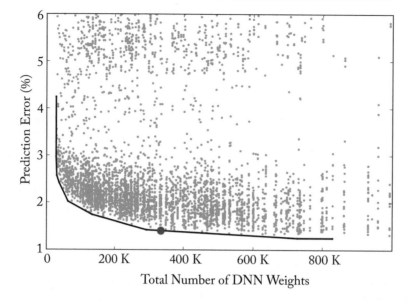

Figure 4.3: Each point is a uniquely trained MNIST neural network. The black line indicates the Pareto frontier, minimizing neural network weights and prediction error. The red dot indicates the chosen network. Parameters swept: 3 to 5 hidden layers, 32 to 512 nodes per layer, and L1/L2 weight parameters from 0 to 10^{-6}.

By conducting a sweep, the user has an opportunity to trade off, at a fine-grain, the initial network size and error. For example, a neural network with three hidden layers and 256 nodes per layer requires roughly 1.3 MB of storage, while a network with 512 nodes per layer requires 3.6 MB (assuming 4 byte weights). This 2.8× storage increase only improves absolute model accuracy 0.05%. The red dot in Figure 4.3 corresponds to an optimal network that balances memory requirements versus incremental accuracy improvements. The resulting network has a prediction error of 1.4%. See selected neural network hyperparameter settings for MNIST and other datasets in Table 4.1.

4.3.2   ACCELERATOR DESIGN SPACE

The second stage of Minerva takes the selected and trained neural network from Stage 1 and searches the accelerator microarchitectural design space for high-performing designs. Evaluating the accelerator design space entails generating and evaluating thousands of unique accelerator implementations, all capable of executing the same neural network. To explore the space efficiently, Minerva uses Aladdin and ultimately yields a power-performance Pareto frontier. From the frontier, users are free to select whichever design meets their specifications. Note that because Minerva reduces power by around an order of magnitude, a user may want to select a

Table 4.1: Application datasets, hyperparameters, and prediction error

Dataset			Hyperparameters				Error (%)		
Name	Inputs	Outputs	Topology	Params	L1	L2	Literature	Minerva	σ
							139		
Forest	54	8	128×512×128	139 K	0	10^{-2}	29.42 [11]	28.87	2.7
							16		
WebKB	3418	4	128×32×128	446 K	10^{-6}	10^{-2}	14.18 [16]	9.89	0.71
							16		

point above their power budget knowing that the optimizations will bring it down to maximize performance. This design is taken as the baseline and is indicative of how neural network accelerators are built today. Note that all presented efficiency gains are with respect to this baseline, not a CPU or GPU. We feel that, as a large body of related work already exists outlining the benefits of specialized hardware for machine learning, using an accelerator as a baseline makes more sense. This is especially true with respect to the constrained embedded and IoT devices, which are largely the target of the Minerva optimizations.

The high-level neural network accelerator architecture is shown in Figure 4.4a, representing the machine specified by the description fed to Aladdin. The accelerator consists of memories for input vectors, weights, and activations as well as multiple datapath lanes that perform neuron computations. Figure 4.5 shows the layout of a single datapath lane, consisting of two operand fetch stages (**F1** and **F2**), a MAC stage (**M**), a linear rectifier activation function unit (**A**), and an activation writeback (**WB**) stage. The parts in black are optimized by Aladdin to consider different combinations of intra-neuron parallelism (i.e., parallelism and pipelining of per-neuron calculations), internal SRAM bandwidth for weights and activations (**F1** and **F2**), and the number of parallel MAC operations (**M**). These features, in addition to inter-neuron parallelism (i.e., how many neurons are computed in parallel), collectively describe a single design point in the design space shown in Figure 4.4b. The red wires and boxes in Figure 4.5 denote additional logic needed to accommodate the Minerva optimizations.

The results of the design space exploration are shown in Figure 4.4b. Each point in the figure is a unique accelerator design, and the red dots indicate the design points along the power-performance Pareto frontier. The area and energy consumed by each of these Pareto design points is further shown in Figure 4.4c. The neural network kernel is embarrassingly parallel within a single layer, and the bottleneck to performance- and energy-scaling quickly becomes memory bandwidth. To supply bandwidth, the SRAMs must be heavily partitioned into smaller memories, but once the minimum SRAM design granularity is reached, additional partitioning becomes wasteful as there is more total capacity than data to store. This scaling effect levies a heavy area penalty against excessively parallel designs and provides little significant energy

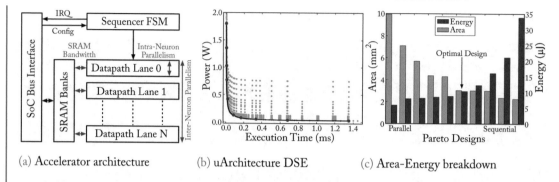

(a) Accelerator architecture   (b) uArchitecture DSE   (c) Area-Energy breakdown

Figure 4.4: (a) A high-level description of our accelerator architecture, (b) results from a DSE over the accelerator implementation space, and (c) energy and area analysis of resulting Pareto frontier designs.

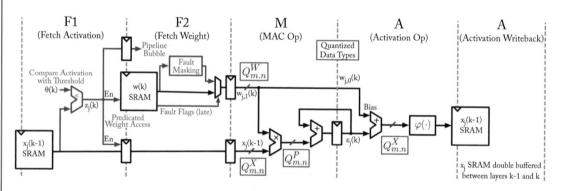

Figure 4.5: The microarchitecture of a single datapath lane. Modifications needed for optimizations are shown in red.

improvement, as seen in the most parallel designs, on the left side of Figure 4.4c. The baseline, indicated as the "Optimal Design" in Figure 4.4c, is a balance between the steep area increase from excessive SRAM partitioning versus the energy reduction of parallel hardware. Under these constraints, the chosen design maximizes performance and minimizes power.

This baseline design is taken to be representative of how neural network accelerators are built today, considering only safe optimizations. There are additional safe optimization tricks (e.g., input batching for increased locality) as well as different architectures (e.g., dataflow [23]). These optimizations are also amenable to unsafe optimizations and could be considered by modeling them in Minerva.

4.4   LOW-POWER NEURAL NETWORK ACCELERATORS: UNSAFE OPTIMIZATIONS

We now apply unsafe optimizations to our baseline accelerator. First, we need to bound the model error by computing the ITN.

Bounding Prediction Error   Minerva modifies the calculations performed by the original neural network in order to optimize power and chip area for the resulting hardware accelerator. This comes with the possibility of a small increase in prediction error. To maintain neural network accuracy, we constrain the cumulative error increase from all Minerva optimizations to be smaller than the intrinsic variation of the training process. This interval is not deterministic, but sensitive to randomness from both the initialization of the pretraining weights and the stochastic gradient descent (SGD) algorithm. This is an example of how we compute the ITN metric.

Figure 4.1 shows the average prediction error and a corresponding confidence interval, denoted by ± 1 standard deviations, obtained across 50 unique training runs of networks with the same hyperparameters. We use these confidence intervals to determine the acceptable upper bound on prediction error increase due to Minerva optimizations. For MNIST, the interval is $\pm 0.14\%$. Table 4.1 enumerates the intervals for the other datasets. As mentioned before, we are not interested in sacrificing accuracy. If the application domain is tolerant to some less accurate model, ITN is still a useful metric to report as it provides context as to how much the model is being approximated in a rigorous way.

4.4.1   DATA TYPE QUANTIZATION

Stage 3 of Minerva aggressively optimizes neural network bitwidths. The use of optimized data types is a key advantage that allows accelerators to achieve better computational efficiency than general-purpose programmable machines. Prior work demonstrates that neural networks are amenable to fixed-point data types (see the survey for a thorough treatment of related work). Minerva offers researchers a method to understand the extent to which neural network variables can be quantized and automatically quantifies the bits-per-type vs. model accuracy trade-off. Additionally, we note that while many choose to use a single, global data type for design simplicity, this is an unnecessary and conservative assumption, resulting in data type bloat and inefficiencies. The use of fine-grained, per-type, per-layer optimizations significantly reduce power and resource demands.

Fixed-Point Datapath Design   Figure 4.5 identifies three distinct signals that Minerva quantizes independently: $x_j(k-1)$, neuron activity (SRAM); $w_{j,i}(k)$, the network's weights (SRAM); and the product $w_{j,i}(k) \cdot x_j(k-1)$, an intermediate value that determines the multiplier width. The notation $Q_{m.n}$ describes a fixed-point type of m integer bits (including the sign bit) and n fractional bits (i.e., bits to the right of the binary point). For each independent signal, we consider a separate fixed-point representation, named accordingly: $Q_{m.n}^{X}$, $Q_{m.n}^{W}$, and $Q_{m.n}^{P}$.

Optimized Fixed-Point Bitwidths   The conventional approach to using fixed-point types in neural networks is to select a single, convenient bitwidth (typically 16 bits [18, 88]) and verify that the prediction error is within some margin of the original floating-point implementation. To improve upon this approach, Minerva considers all possible combinations and granulates of m and n for each signal within each network layer, independently.

Figure 4.6 presents the resulting number of bits $(m + n)$ from fine-grained bitwidth tuning for each signal on a per-layer basis for MNIST. The required number of bits is computed by iteratively reducing the width, starting from a baseline type of $Q_{6.10}$ for all signals (shown in Figure 4.6 as Baseline), and running the Keras SW models to measure prediction error. The minimum number of bits required is set to be the point at which reducing the precision by 1 bit (in either the integer or fraction) exceeds MNIST's error confidence interval of $\pm 0.14\%$.

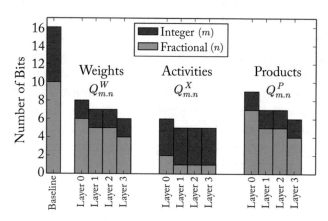

Figure 4.6: Minimum precision requirements for each datapath signal while preserving model accuracy within our established error bound.

The datapath lanes process the neural network graph sequentially, in a time-multiplexed fashion. Hence, even though per-layer optimization provides the potential for further width reductions, all datapath types are set to the largest per-type requirement (e.g., all weights are 8 bits). While this may seem wasteful, it is in fact the optimal design decision. With respect to weight SRAM word sizes, the savings from removing one to two additional bits saves 11% power and 15% area but requires a larger number of unique SRAMs. Instantiating two different word-sized SRAMs tailored to each layer, as opposed to having a single 8 bit word SRAM, results in a 19% increase in area. In a similar vein, it is difficult to modify the MAC stage logic to allow such fine-grained reconfiguration without incurring additional overheads.

Using per-type values of $Q_{2.6}^{W}$, $Q_{2.4}^{X}$, and $Q_{2.7}^{P}$ provides a power saving of 1.6× for MNIST and an average of 1.5× across all datasets compared to a traditional, global, fixed-point type of $Q_{6.10}$ as used in [18].

4.4.2   SELECTIVE OPERATION PRUNING

Stage 4 of Minerva reduces the number of edges that must be processed in the dataflow graph. Using empirical analysis of neuron activity, we show that by eliminating operations involving small activity values, the number of weight read and MAC operations can be drastically reduced without impacting prediction accuracy.

Analysis of Neural Network Activations   Figure 4.7 shows a histogram of neuron activations over all of the test vectors for the MNIST dataset. Immediately visible is the overwhelming number of zero- and near-zero values. Mathematically, zero-valued activations are ineffective: guaranteed to have no impact on prediction error. Skipping operations involving these values can save power both from avoided multiplications and SRAM accesses. Intuitively, it also stands to reason that many near-zero values have no measurable impact on prediction error. Thus, if we loosen the definition of "insignificant" to encompass this large population of small values, we stand to gain additional power savings with no accuracy penalty.

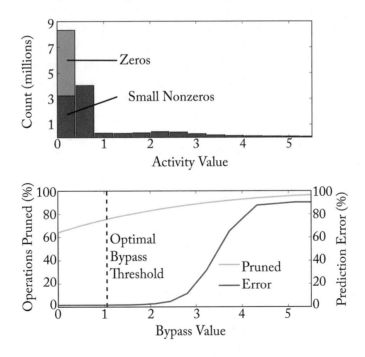

Figure 4.7: Analysis of neuron activations and sensitivity of prediction error to pruning.

Stage 4 quantifies the relationship between activity value and overall error. Our software model elides operations involving any neuron activity below a given threshold and evaluates the network as if those activations were zero. The resulting curves in Figure 4.7 show that even if we remove all activations with a magnitude of less than 1.05, the overall prediction error is unaffected. The green pruned-operation curve tracks the cumulative number of activations smaller than a given value. A threshold value of 1.05 is larger than approximately 75% of activations, meaning we can safely prune three out of every four MAC and SRAM read operations.

While this may sound surprising, there is some intuition behind this phenomenon. Rectifier activation functions eliminate negative products, so we should expect approximately half of the internal activity values to be zero from this effect alone. This is largely responsible for the high number of operations pruned even with a threshold close to zero (the y-intercept of the pruned-operations curve in Figure 4.7). Additionally, neural networks based on rectifier activation functions are known to grow increasingly sparse with deeper topologies as a result of successive decimation [53]. The combination of these effects results in a remarkable number of insignificant operations. In addition to quantization, selective pruning further reduces power 1.9× for MNIST and an average of 2.0× over all datasets.

Predicating on Insignificant Operations   The sparsity patterns in the activations is a dynamic function of the input data vector. As such, it is not possible to determine *a priori* which operations can be pruned. Therefore, it is necessary to inspect the neuron activations as they are read from SRAM and dynamically predicate MAC operations for small values. To achieve this, the datapath lane (Figure 4.5) splits the fetch operations over two stages. **F1** reads the current neuron activation $(x_j(k-1))$ from SRAM and compares it to the per-layer threshold $(\theta(k))$ to generate a flag bit, $z_j(k)$, which indicates if the operation can be skipped. Subsequently, **F2** uses the $z_j(k)$ flag to predicate the SRAM weight read $(W_{ij}(k))$ and stall the following MAC (**M**), using clock-gating to reduce the dynamic power of the datapath lane. The hardware overhead for splitting the fetch operations, an additional pipeline stage, and a comparator, are negligible.

4.4.3   SRAM FAULT MITIGATION

The final stage of Minerva optimizes SRAM power by reducing the supply voltage. While the power savings are significant, lowering SRAM voltages causes an exponential increase in the bitcell fault rate. We only consider reducing SRAM voltages, and not the datapath, as they account for the vast majority of the remaining accelerator power. To enable robust SRAM scaling, we present novel codesigned fault-mitigation techniques.

SRAM Supply Voltage Scaling   Scaling SRAM voltages is challenging due to the low noise margin circuits used in SRAMs, including ratioed logic in the bitcell, domino logic in the bitline operation, and various self-timed circuits. Figure 4.8 shows SPICE simulation results for SRAM power consumption and corresponding bitcell fault rates when scaling the supply voltage of a 16 KB SRAM array in 40 nm CMOS. The fault rate curve indicates the probability of a single bit error in the SRAM array. This data was generated using Monte Carlo simulation with 10,000 samples at each voltage step to model process variation effects, similar to the analysis in [3] but with a modern technology.

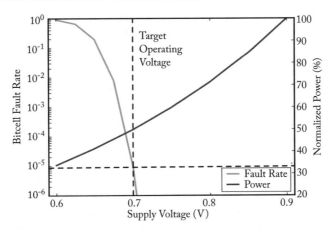

Figure 4.8: SRAM supply voltage scaling trends for fault rate and power dissipation.

SRAM power decreases quadratically as voltage scales down while the probability of any bitcell failing increases exponentially. If we target an operating voltage of 0.7 V, we could approximately halve the power consumption of the SRAM and, according to Figure 4.8, operate with seemingly negligible fault probability. However, many practical effects make this fault rate far higher than the ideal case, including process variation, voltage noise, temperature, and aging effects. As a result, it is important to be able to tolerate fault rates beyond this point.

SRAM Fault Detection   A number of techniques have been described to detect faults arising from voltage scaling, including parity bits [77], Razor double-sampling [38, 86], Razor transition-detection [14], and various canary circuits [113]. A single parity bit enables fault detection simply by inspecting the read SRAM data. In contrast, Razor and canary circuits provide fault detection by monitoring delays in the circuits. None of these solutions correct suspected-faulty SRAM data but instead indicate when a fault may have occurred. The overheads of these fault detection methods vary relative to the protection provided. Parity can only detect an odd number of errors and provides no information as to which bit(s) in the word the fault(s) may have affected. In the case of neural network accelerators, the word sizes are relatively small (Section 4.4.1); the overhead of a single parity bit is approximately 11% area and 9% power (compared to the nominal voltage). Thus, anything more than a single bit is prohibitive.

In this work, we instead employ the Razor double-sampling method for fault detection. Unlike parity, Razor monitors each column of the array individually, which provides two key advantages: (1) there is no limit on the number of faults that can be detected, and (2) information is available on which bit(s) are affected. The relative overheads for Razor SRAM fault detection are modeled in our single-port weight arrays as 12.8% and 0.3% for power and area respectively [86]. This overhead is modeled in the SRAM characterization used by Aladdin.

Mitigating Faults in Neural Networks   Razor SRAMs provide fault detection, not correction. Correction mechanisms require additional consideration and are built on top of detection techniques. In fault-tolerant CPU applications, a typical correction mechanism is to recompute using checkpoint and replay [13, 14, 39], a capability specific to the sophisticated control-plane logic of a CPU employing pipeline speculation. In an accelerator, implementing replay mechanisms incurs significant, and in this case unnecessary, overheads. To avoid this, we use a lightweight approach that does not reproduce the original data but instead attempts to mitigate the impact of intermittent bit-flips to the neural network's model accuracy. This is reminiscent of an approach studied for fault-tolerant DSP accelerators [143].

We extend the Keras neural network model to study the impact of SRAM faults incurred from low-voltage operation on prediction accuracy. Faults are modeled as random bit-flips in the weight matrix (**w**). Figure 4.9 shows how these faults result in degraded prediction accuracy as an exponential function of fault probability. Figure 4.9a shows weight fault tolerance without any protection mechanism; here, even at relatively small fault probabilities ($< 10^{-4}$), the prediction error exceeds our established error bound of ITN 1. Once the fault rate goes above 10^{-3}, the model is completely random with 90% prediction error. This result intuitively follows from the observations regarding the sparsity of the network activity in the previous section—if the majority of neuron outputs are zero, a fault that flips a high-order bit can have a catastrophic affect on the classification.

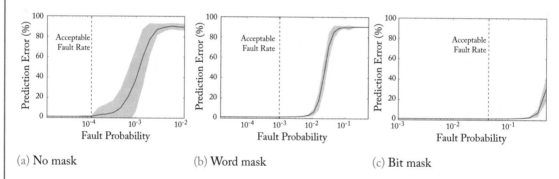

(a) No mask   (b) Word mask   (c) Bit mask

Figure 4.9: The sensitivity of weight matrices to proposed fault-mitigation techniques. The figures show the impact faults have on accuracy when mitigating faults with (a) no mask, (b) word masking, and (c) bit masking. The vertical dashed lines correspond to the maximum tolerable fault rate that satisfies our 0.14% absolute increase to prediction error bound.

To prevent prediction accuracy degradation, we combine Razor fault detection with mechanisms to *mask* data toward zero when faults are detected at the circuit level. Masking can be performed at two different granularities: *word masking* (when a fault is detected, all the bits of the word are set to zero) and *bit masking* (any bits that experience faults are replaced with the sign bit). This achieves a similar effect to rounding the bit position toward zero. Figure 4.10 gives a simple illustration of word-masking and bit-masking fault mitigation.

Figure 4.10 illustrates the benefits of masking. Masking the whole word to zero on a fault is equivalent to removing an edge from the neural network graph, preventing fault propagation. However, there is an associated second-order effect of neuron activity being inhibited by word masking when a fault coincides with a large activity in the preceding layer. Compared to no protection, word masking is able to tolerate an order of magnitude more bitcell faults.

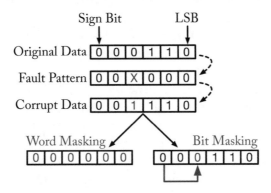

Figure 4.10: Illustration of word-masking (faulty weights set to zero) and bit-masking (faulty bits set to sign bit) fault-mitigation techniques.

By masking only the affected bit(s) rather than the whole word, bit masking (Figure 4.9c) is the strongest solution. Bit masking limits error propagation through the fully connected network of a neural network, and its bias toward zero complements the natural sparsity of ReLU-activated networks. The combination of Razor fault detection and bit-masking fault mitigation allows the weight SRAMs to tolerate 44× more faults than word masking. With bit masking, 4.4% of SRAM bitcells can fault without loss of prediction error. This level of fault tolerance allows us to confidently drop our SRAM supply voltage by more than 200 mV, reducing overall power consumption of the accelerator by an additional 2.7× on average and 2.5× for MNIST.

HW Modifications for Mitigation   Bit masking requires modifications to the weight fetch stage (**F2**) of the datapath lane, as illustrated in Figure 4.5. The Razor SRAM provides flags to indicate a potential fault for each column of the array, which are used to mask bits in the data, replacing them with the sign bit using a row of two-input multiplexers. The flag bits are prone to metastability because (by definition) timing will not be met in the case of a fault. However, metastability is not a significant concern as the error flags only fan out into datapath logic, as opposed to control logic, which can be problematic. The flags are also somewhat late to arrive, which is accommodated by placing the multiplexers at the end of the **F2** stage (Figure 4.5).

4.5 DISCUSSION

Up to this point, we have focused only on MNIST for clarity in the presentation of the Minerva design flow and optimizations. In this section, we demonstrate the generality of Minerva by considering four additional datasets; each is run through the Minerva design flow described in Sections 4.3.1 to 4.4.3.

Datasets Evaluated   To demonstrate the generality of Minerva and the unsafe optimizations considered, we consider four additional classification datasets:[2] (1) *MNIST*: images of hand-written digits from 0 to 9 [90]; (2) *Forest*: cartographic observations for classifying the forest cover type [11]; (3) *Reuters*-21578 (Distribution 1.0): news articles for text categorization [16, 92]; (4) *WebKB*: web pages from different universities [36]; (5) *20NG*: newsgroup posts for text classification and text clustering [79].

Figure 4.11 presents the resulting accelerator designs, indicating the power reduction of each with respect to each optimization stage. On average, Minerva-generated neural network accelerators dissipate 8.1× less power, operating at the 10 s rather than 100 s of mWs. While the overall power reduction for each dataset is similar, the relative benefits from each optimization differ. For example, MNIST benefits more from quantization (1.6×) than WebKB (1.5×), while WebKB is far more amenable to operation pruning—2.3× vs. MNIST's 1.9×. This is caused by differences in the application domain and input vectors.

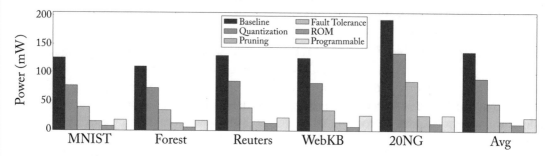

Figure 4.11: Results from applying the Minerva design flow to five application datasets to investigate generality. Each successive optimization ensures that the compounding error does not exceed the established threshold. The bars labeled ROM show the benefit of full customization, wherein weights are stored in ROMs rather than SRAMs. Those that are labeled "Programmable" show the overhead each dataset incurs when an accelerator is designed to handle all five datasets.

Balancing Specialization and Flexibility   With Minerva, we are able to consider the trade-offs of building a fully optimized accelerator capable of performing only one function against

---

[2]When choosing these datasets, Fathom was not available. We chose datasets we believed best represented a broad class of problems of interest to the IoT and embedded space.

building a programmable accelerator able to run any of the datasets using our assumed network topologies. In Figure 4.11, the bars labeled ROM indicate the power saved when the weights are stored using ROM instead of SRAM. This additional optimization further reduces power on average by 1.9×.

We also consider building a configurable accelerator. To accommodate all datasets, the configurable accelerator's parameters are set to the maximum of each individually optimized design (i.e., it supports 20NG's 21979 input size and up to $256 \times 512 \times 512$ nodes per layer). This design consumes an average of 24mW across all datasets and uses 1.4× and 2.6× more power than dataset-specific SRAM and ROM implementations respectively. The largest overhead introduced by the configurable design relative to building individual accelerators for each dataset is due to memory leakage. With selective pruning, the dynamic power remains minimal. The inefficiencies of generality, compared to an accelerator tuned to each individual dataset, are reflected in the bars labeled "Programmable" in Figure 4.11.

4.6   LOOKING FORWARD

In this chapter, we presented a methodology for designing and optimizing neural network accelerators. The Minerva flow is a particular realization of the more general approach necessary to building machine learning accelerators looking forward. As the compute cost of these models continues to grow, and their usage also increases, more aggressive optimizations will be needed.

As more unsafe optimizations are proposed, it becomes important to understand how optimizations interact with each other. For example, the fixed-point precision used has a strong effect on SRAM fault tolerance. If too many bits are used for the integer representation, then the model can tolerate far fewer faults in weights as the fault impact (the delta of the faulty and correct weight) is large. This is because having more integer bits increases the probability that a fault results in an error in a large magnitude bit. As more optimizations are developed, this understanding will be vital for proper designs.

While Minerva provides a solution for many immediate problems with neural network accelerator design, it also creates some that demand attention. The first is how one goes about optimizing all the free parameters, which are often correlated (as mentioned above). More optimizations means more design space dimensions, and as it stands, even the first two safe Minerva stages are challenging to explore thoroughly. To minimize run times, Minerva optimizations are applied sequentially; this is suboptimal, however, as ideally all parameters would be explored and tuned together. Accomplishing this goal requires more intelligent means of exploring design space. While Minerva currently conducts a grid-search, this explorations could be optimized with tools such as those presented in [67, 91, 115, 127].

Another open question is this: which unsafe optimizations can be built into the training process? By informing loss functions about model properties that enable efficient execution, it should be possible to explicitly train efficient networks rather than always extracting implicit redundancy.

CHAPTER 5

A Literature Survey and Review

5.1 INTRODUCTION

In recent years, there has been a phenomenal surge in publications on hardware for neural networks in both the computer architecture and circuits venues. This chapter is intended to provide a survey and review of research on computer architecture for machine learning, as the field presently stands. As we saw in Chapter 1, historical improvements in hardware performance have led to a virtuous cycle generating significant innovations in neural network research. Therefore, as computer architects begin to make inroads to increasing performance for neural networks, we expect the rate of progress in the field to continue to grow. The most relevant material is currently spread between the overlapping topics, and associated publication venues, of machine learning theory (i.e., algorithms), computer architecture, and circuits. Here we attempt to draw parallel threads across these sometimes disparate fields.

The machine learning community also tends to champion modern research dissemination methods, such as extensively publishing preprints using `arxiv.org`, public, open online review of conference submissions, accessible and insightful blog posting, and extensive open sourcing of reference software implementations. The sheer rate of progress seen may be in part attributed to this, as it removes barriers and encourages experimentation, spurring on innovations and very rapid publication cycles. A significant challenge with this rate of progress is that inevitably, much of this research will date rapidly. Unfortunately, the same is probably true of this survey chapter, so we hope this forms a useful basis to start research, even as the topic evolves in the near future.

5.2 TAXONOMY

The material in this chapter is organized using a simple taxonomy outlined in Table 5.1. The main axis for the taxonomy is the familiar computer architecture stack consisting of three levels of abstraction: algorithms, architecture, and circuits. The orthogonal secondary axis consists of a number of fundamental research threads that cut across the stack: data types, model sparsity, model support, data movement, and fault tolerance. The remainder of this chapter is structured to loosely follow this axis. The three abstraction levels form major subsections of this chapter, with the research threads at each level of the stack discussed within each subsection. Because of

Table 5.1: Taxonomy of surveyed literature, arranged by the orthogonal axes of computer architecture abstraction stack and research theme. See the text for a description of each research theme.

	Algorithms	Architecture	Circuits
Data Types	Gupta et al. [58] Anwar et al. [6] Lin et al. [94] Lin et al. [96] *BinaryConnect* [35] *BinaryNet* [34] 75 21 20 126 31 43 59 70	22 109 4 114 144 61	
Model Support		*PuDianNao* [97] *Cambricon* [98] *TPU* [80] *Catapult* [17] *DNN Weaver* [124] *Tabla* [101] 80 111 22 5 116	93 122 27 12 82
Fault Tolerance			Temam et al. [137] Du et al. [44] Kung et al. [87] *DNN Engine* [144]

this, some papers appear more than once, focusing on different aspects of the work. Also, some subsections do not include every secondary taxon.

The research threads we focus on were chosen to represent the most prevalent in the contemporary literature. They range from very practical considerations, such as the choice of data types, to more academic topics such as various fault tolerance themes that have been published.

- **Data types** offer significant scope for optimization of specialized hardware. Neural networks generally involve a huge number of numeric operations, and therefore the size of the data types determine power and performance limits. However, overly optimizing data types can limit model support.

- **Model sparsity** is an inherent property in many neural networks and is widely believed to be a necessary condition for the process of learning. Typically, sparsity exists in (1) input data, (2) model parameters, and (3) intermediate activation data. Exploiting data sparsity leads to compelling improvements in power and performance but introduces complexity.

- **Model support** refers to the challenge of supporting a range of neural network models efficiently in hardware. The trade-off between efficiency and specialization is an interesting topic for computer architects and deserves attention in the area of machine learning. Beyond inference, hardware support for training is also starting to gain attention.

- **Data movement** is also a significant challenge. In particular, the common fully connected classifier layer requires parameterization that is quadratic with respect to the size of the layer. In contrast, convolutional layers often involve working with many parallel feature maps, which can add up to a huge number of data items. In both cases, moving neural network data is an expensive and challenging task.

- **Fault tolerance** is a property of many machine learning systems during both inference and training and is intuitively known from the biological metaphor. Fault tolerance is a very desirable property for computer architectures, and hence this has received attention from architects.

5.3   ALGORITHMS

It is often said that the biggest gains come from optimizations at the top of the stack. To emphasize this, we start by surveying some key developments in neural network algorithms and techniques. Significant progress is being made in many challenging areas in machine learning. Here we highlight some papers from the machine learning community that are pertinent to hardware design. The main goal in machine learning is to demonstrate techniques that lower test error on a given dataset. One of the key techniques used to achieve progress in this regard is to train larger models. This, in itself, is a goal that presents a number of challenges with respect to training. It also incurs ever-increasing compute and memory overheads. This fact is not lost on machine learning researchers, who are in fact heavily motivated to make their models more efficient as this in turn allows larger models to be trained in a practical time frame, using commodity hardware. Therefore, the machine learning papers presented in this section focus on optimizing data types and exploiting sparsity. These techniques may be familiar to experienced architects working in other domains.

5.3.1 DATA TYPES

The optimization of numerical data types is closely related to model compression. In both cases, the goal is to reduce the size of the data required to compute the neural network. The benefit of this is that the time and energy associated with moving data around is diminished. Depending on the reduction achieved, it may also result in the data residing at a higher level of the memory hierarchy, which again saves energy and reduces access latencies. A particularly important threshold in regard to the memory system is whether the data can fit in on-chip memory. This is an important tipping point, because off-chip DRAM access is orders of magnitude more expensive.

Reducing numerical dynamic range and/or precision leads to a reduced storage requirement as less bits are needed to represent each value. It also reduces the power and latency of the arithmetic operations themselves. However, this eventually impacts the test accuracy of a neural network model. A process known as quantization is typically used to transform a model trained using floating-point numbers into narrow fixed-point types. Although small fixed-point data types have been popular, there are many other options typically based around variants and hybrids of otherwise well-known fixed-point and floating-point number systems. Optimizing data types is probably one of the most popular techniques, and many GPUs now natively support small fixed-point types and floating-point for this purpose.

A number of papers have previously considered optimizing data types for inference with larger convolutional models [6, 94]. Both present approaches to quantize data types heavily, often with the goal of running these models on mobile devices that have more constrained storage. Exemplary results on CIFAR-10 show a reduction of over 20% in model size with no loss in accuracy [94]. Google's *TPU* [80] also emphasizes the use of small 8-bit fixed-point types by default. The *TensorFlow* framework is used to quantize neural networks to 8-bit values, which are then used on the TPU hardware to achieve improved throughput and energy efficiency. Eight-bit fixed-point multiplication can have 6× less energy and 6× less area than the equivalent IEEE 754 16-bit floating-point operation, and for addition, 8-bit fixed-point is 13× lower in energy and 38× less in area [80].

Recently, it has even been shown that weights can be represented with a single bit per parameter without degrading classification performance on popular datasets such as MNIST, CIFAR-10, and SVHN [34, 35, 96]. The advantage of being particularly aggressive in this way is the strength reduction of multiplication to a simple manipulation of the sign bit. In fact, it was also observed that accuracy can improve due to a stochastic regularization effect that can be related to dropout schemes [69]. Improvements in training techniques, specifically batch normalization [76], was probably needed to enable these so-called BinaryConnect networks, as normalization circumvents potential problems with scale in weights. Ternary weights have also been demonstrated [134] and were again found to have negligible accuracy loss, under the condition that the network is "large," i.e., limited by training data and not the number of free parameters.

Concerning recurrent networks specifically, binarization, ternarization, and a power-of-2 representation have all been examined experimentally in [110]. Given the more involved training methods necessary with recurrent networks, it is perhaps surprising that many of these approaches still work well. An exception was found with GRUs, where the authors were unable to binarize the weights of the network and still maintain functionality.

Considering training, it has been shown that neural networks can be trained with low-precision fixed-point arithmetic, provided care is taken with the rounding technique, which should be stochastic [58]. Results on MNIST and CIFAR-10 datasets show that deep networks can be trained using 16-bit fixed-point data types, with negligible degradation in classification accuracy. This paper also highlights that attention must be paid to important details, such as the use of stochastic rounding of values when accumulating large dot products. Nonetheless, except for some results on smaller-scale datasets, it seems that in general the use of more aggressive quantization in the training process is still an open research topic, and typically 16-bit floating-point values are much more common.

5.3.2   MODEL SPARSITY

An identifying feature of neural network models is the scale of the model parameterization. Clearly, minimizing the size of the storage required for the network weights can only increase performance and reduce power. This is an advantage in both hardware and software implementations for training and inference. Note that the practical desire to minimize the model size is in stark contrast to the driving goal of deep learning, which is fundamentally to improve accuracy by developing techniques to train larger models. However, some recent work from the machine learning community suggests the tension between model size, performance, and accuracy does not have to be a rigid dichotomy.

Fully connected classifier layers are one of the biggest culprits for inflating model size. The number of parameters in a fully connected layer is a quadratic function of the number of nodes in the layer. Recursive layers suffer with a similar problem since they also require dense matrix-vector operations. However, recursive layers also require parameters for the state from previous time steps. For example, this can introduce a factor of 2× for a unidirectional vanilla RNN layer or 3× for a bidirectional layer. More sophisticated recursive units such as LSTMs demand even larger parameterization. The filter kernels used for convolutional layers are smaller than the input layer and typically are as small as 3×3 or 5×5 elements. Nonetheless, even these can be problematic in big models due to the sheer number of kernels.

At the algorithm level, efficient hyperparameter optimization should be used to minimize the fundamental model size. The network architecture is the most fundamental consideration in this regard. Network choices are mainly driven by intuition and experimentation. Traditionally, fully connected classifiers are employed at the end of many network architectures. Some recently published networks in fact avoid large classifiers at the end of the network [75, 95]. In the case of [95], this is demonstrated while achieving state-of-the-art accuracy. It remains to be seen if this will be an enduring trend. Subsequent detailed tuning of hyperparameters is traditionally considered something of a black art. However, recent work on the systematic application of Bayesian optimization to the problem demonstrates an automated approach [127].

Many in the machine learning community have picked up on the practical advantages of compact models. After all, small models are faster to train and experiment with, which is an advantage to machine learning researchers, who are perhaps otherwise ambivalent about implementation performance. Some alternative models have been proposed with the goal of compact parameterization. The HashedNets model uses the concept of a virtual weights matrix in combination with an indexing hash function to replace the conventional weights matrix (Figure 5.1). The virtual weights matrix is much smaller than the matrix it replaces and therefore reduces the model size significantly. In [21], the concept is proposed and demonstrated on fully connected DNNs. Results show that HashedNets can achieve significant reductions in model size on datasets such as MNIST with negligible impact on test error. The paper also presents the dual of this problem, by fixing the number of parameters (i.e., real weights) and inflating the virtual weight matrix size. This insightful experiment resulted in a 50% reduction in test error for an $8\times$ expansion in the virtual weights, without increasing the required storage for real weights. The concept was further extended to convolutional networks, by performing the weight sharing in the frequency domain [20]. In both cases, training is essentially unchanged (beyond adding the hash function to map virtual weights to real weights), and uses conventional back- propagation.

The weights of neural network models are well-known to be sparse. Regularization techniques essentially enforce a degree of sparsity to prevent overfitting. Fundamentally, it may be that sparsity is in fact a prerequisite for learning, as most models begin overparameterized and unnecessary parameters fall out during the training process. Nonetheless, there appears to be little known about how sparsity relates to other aspects of neural networks. It has been demonstrated that it is possible to learn weight matrices with structured sparsity, such as a Toeplitz matrix [126] structure. The structured matrices can be exploited to implement faster linear algebra operations, such as matrix-vector multiplication, inversion, and factorization. On a keyword-spotting application, compression of $3.5\times$ was achieved at state-of-the-art accuracy.

SqueezeNet [75] is a CNN architecture that focuses on minimizing model size. This is achieved primarily by focusing on small 1×1 filters for convolutions (Figure 5.2) as these are most size efficient. Other concessions to size include decreasing the number of input channels to wide filters and downsampling late in the network. Fully connected layers are also avoided for

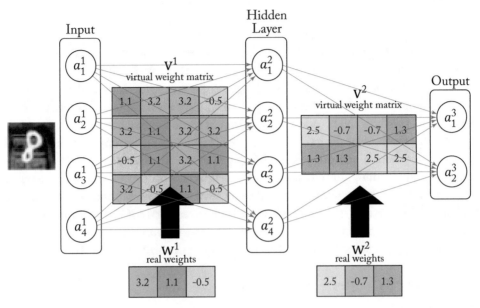

Figure 5.1: *HashedNets* [21] implements DNNs using random weight sharing. This example uses a real weight matrix with a compression factor of $\frac{1}{4}$, which is expanded into a virtual weight matrix. The colors represent the virtual matrix elements that share the same real weight value.

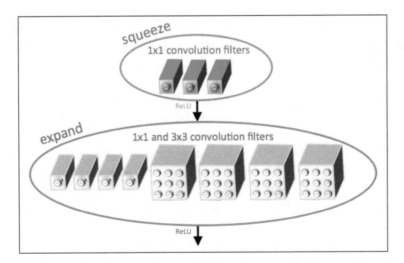

Figure 5.2: *SqueezeNet* [75] uses layers containing small 1×1 convolution kernels, which results in a network achieving similar accuracy to Alexnet but with around 50× fewer parameters.

compactness. Results demonstrate AlexNet [85] accuracy on the ImageNet dataset, with 50×
fewer parameters.

The memory footprint of a conventional model can also be reduced using a postprocessing
technique. Random Edge Removal (RER) [31] is one example, where a fraction of the weights
are randomly set to zero (and can therefore be ignored). Low-Rank Decomposition [43] is
another that has been proposed. Dark Knowledge [70] is an approach that allows more compact
models to be trained from the soft-max outputs of a larger (uncompressed) model. At the time
of writing, [126] provides the broadest comparison of these approaches.

A practical example of the gains that can be achieved using compression on real problems
is given in [60], where a compressed model is generated in three steps. First, a competitive
network is trained. Second, unimportant (small) weights are pruned to reduce the network size.
Third, the pruned network is retrained to tune the remaining weights and recover any accuracy
loss. Following this method, the number of parameters required for Alexnet can be reduced by
9× without incurring accuracy loss. Follow-up work, *Deep Compression* [59], adds clustering and
Huffman encoding and demonstrates a 49× reduction in the size of VGG-16.

5.4   ARCHITECTURE

Neural networks represent a really interesting problem space for architecture specialization. In
essence, the majority of deep learning methods rely heavily on large linear algebra kernels, which
historically have tended to lend themselves to a number of effective algorithm optimizations,
as evidenced by the evolution of BLAS. Compared to many other modern workloads, these
kernels also have some very favorable properties, such as huge data-level parallelism, flexibility
in operation ordering and scheduling, and essentially no data-dependent sequencing. However,
the emphasis on these large linear algebra kernels, with their favorable properties for computer
hardware may change rapidly in the near future as neural networks evolve. Therefore, it may take
a long time for the field to mature to the point where we can understand which computer architecture specialization techniques are most effective. Hence, much of the research work surveyed
here is probably a long way from use in commercial products. Nonetheless, we are starting to
see some evidence of commercial hardware tailored to machine learning, such as Google's *Tensor
Processing Unit* (TPU) [80]. The TPU is a bespoke coprocessor-style ASIC to accelerate neural
network inference in commercial datacenter applications. The specific economies of scale associated with datacenter computing are clearly a good fit for a dedicated ASIC solution. It will be
exciting to see similar developments appearing in other computing application domains in the
near future.

For hardware developers looking to optimize for machine learning workloads, there is an
inevitable tension between efficiency and flexibility, which can be hard to navigate. The most
flexible approach to implementing machine learning applications will always be software based.
The huge data-level parallelism inherent to linear algebra is actually often well served by optimized software routines running on contemporary SIMD CPU or GPGPU hardware. In many

ways, a software-focused approach targeting commodity platforms avoids the huge NRE costs of dedicated ASICs, which are also at huge risk of premature obsolescence owing to breakthroughs in machine learning algorithms.

On the other end of the flexibility scale, specialized hardware always provides a performance and efficiency benefit. Specialized hardware could refer to anything from a new instruction in an established architecture to a dedicated hardware accelerator and anything in between. Currently, there is unprecedented interest in dedicated hardware accelerators for neural network inference. This approach does typically provide a significant efficiency advantage for a number of reasons:

- The overhead of fetching and decoding instructions can be often be significantly reduced. Typically, the ISA can be very small and heavily optimized to reduce the size and number of instructions required. In many accelerator designs, instructions are eschewed entirely and instead all sequencing is driven by a (possibly parameterized) FSM controller.

- A custom hardware accelerator microarchitecture can typically be freely designed independent of constraints relating to a heavily constrained ISA. This follows from the previous point about the diminished importance of ISA. Hence, limitations in the design of general purpose programmable hardware are often vastly relaxed: decoding of instructions, ILP windows, complex pipeline hazards, etc.

- Data types employed in the machine can be optimized heavily.

- Tightly coupled memory structures such as scratchpads, caches, FIFOs, and queues can be heavily optimized and freely employed.

Dedicated hardware accelerators do not have to mean ASIC design. For a long time, FPGAs have represented an interesting compromise between the flexibility of software and the performance of ASICs. FPGAs have historically been a somewhat niche solution, but they may be very well suited to machine learning applications, since they have good support for integer arithmetic that is sufficient for neural network inference. They also provide the aforementioned reprogrammability to allow for changes in machine learning algorithms. They are generally not low-power platforms and hence have had little success in mobile platforms to date. But there has been significant interest lately in FPGAs for datacenter applications, as evidenced by the Microsoft *Catapult* [17] effort. It's very likely that neural networks are a good fit for FPGAs in the datacenter, and we expect to see more results in this area in the short term.

In the remainder of this section, we review various architecture proposals that mainly target ASIC implementation. These designs range in programmability from efficient fixed-function designs to programmable machines with optimized ISAs. Architecture research on training seems to lag attention on inference. In many ways, this is not surprising since for commercial deployment, it is the inference performance that we care about. However, improvements in the speed in which we can train models is still important, even if training could be considered more of a one-time cost. Therefore, most of the research discussed here is inference, unless training is specifically mentioned.

5.4.1   MODEL SPARSITY

From the computer architecture perspective, the various forms of data sparsity in neural networks present an interesting opportunity. In Section 5.3, we looked at research on model compression, which is currently a very rapidly moving topic. However, the model parameterization is merely one facet of sparsity that can be exploited. The intermediate activation data flowing through the network, and even the input data vector, can also be sparse. In many cases, it is hard to fully exploit sparsity in software implementations due to the overheads of indirect memory access. However, with hardware accelerators, there exists much more flexibility that can lead to massive savings in compute and memory bandwidth. The potential of such optimizations has not been lost on architects, and a number of papers have explored mechanisms to exploit sparsity.

Sparsity in the input data is perhaps the most intuitive. The vast majority of datasets used in machine learning exhibit significant sparsity. For example, natural images typically have regions that are of a very similar color, e.g., pictures with large areas of blue sky. This can be leveraged in many applications by compressing the data to reduce both the memory footprint and the time and energy required to load the input data. In this way, two recently proposed image recognition CNN accelerators [22, 109] use compression techniques to reduce memory bandwidth, and thus power. In [22], a run-length compression scheme is used, achieving a 2× reduction in average image bandwidth. In contrast, the accelerator in [109] uses a simple Huffman-based compression method and achieves up to 5.8× bandwidth reduction for AlexNet.

Beyond compression of the input data, significant sparsity also exists in the intermediate data (i.e., the activations). The *Minerva* [114] architecture study shows that the activation data that flows through fully connected layers during inference is overwhelmingly dominated by values of small magnitude. This includes both zero and small nonzero values. The position of these values is dynamic, changing for each input vector. Intuitively, an explanation for this is that not all the features learned by the network during training are activated for every input vector, and, therefore, we expect many neurons to be "quiet" during any single prediction. The widely adopted ReLU activation function in modern neural networks results in any negative neuron accumulation generating an exactly zero output.

Since multiplication and addition with one or more zero operands is redundant, we can skip operations involving these values. Skipping redundant operations can reduce energy and increase throughput since less-effective work is performed and there are fewer memory accesses. In addition to this, some of the small nonzero values can also be considered "insignificant" and skipped during inference. A fully connected model trained on the MNIST dataset can tolerate skipping as much as 75% of activation data with negligible increase in test error. This result is then exploited in the microarchitecture by adding a comparator at the front end of the pipeline that thresholds incoming activations and predicates both weight loads and MAC operations.

Skipping as many as three out of four SRAM/MAC operations on fully connected networks was shown to result in a power reduction of 1.9× (MNIST dataset) to 2.3× (WebKB dataset). Further results on this work were given in Chapter 4.

DNN Engine [144] extends this idea with a more efficient architecture for exploiting data sparsity in fully connected layers. The idea is that instead of performing thresholding on the operands before predicating the MAC operation, it can instead be done at the end of the pipeline after the neuron value is completed and before it is stored for use in the next layer (Figure 5.3). Hence, the comparison is performed only once for each neuron after applying the activation function, rather than every time the activation is used in the following layer. When the neuron output is written back, only significant activations that are larger than the threshold need to be stored, which reduces the number of memory writes for activation data. And, correspondingly, when reading in activations from the previous layer, only significant activations are read, therefore the number of cycles is vastly reduced and there are no energy-degrading pipeline bubbles from stalls. Hence, in this approach, the number of operations is reduced as before, but now the number of cycles is also reduced by a similar factor. Therefore, energy savings of 4× and simultaneous throughput improvement of 4× is possible for fully connected MNIST models. Since the stored activations from a complete layer are no longer contiguous, it is necessary to maintain a list of stored activation indexes. This list is then used to generate the corresponding weight indexes. The list of activation indexes is negligible as the maximum number of entries is given by the number of nodes in the layer.

Sparsity has also been studied in much larger models. The *Cnvlutin* [4] architecture study examined data sparsity for large CNN models and found that the activation data in CNN layers is also sparse. The number of exactly zero values ranging from 37% to 50% and averaging 44% across a number of common image classification CNNs, including AlexNet, GoogLeNet and VGG. Two CNN accelerators implemented in silicon [22, 109] make use of a comparator to check the activation data against a threshold and then predicate subsequent operations on the resulting flag, which reduces power, in a similar fashion to *Minerva* in the case of fully connected layers. Even better, *Cnvlutin* makes use of a new sparse data structure that allows activations to be encoded as a value/offset pair. This data structure is incorporated into the front end of the DaDianNao [25] accelerator. Reported simulation results show 1.37× average performance improvement compared to DaDianNao, for a set of image classification CNNs, including AlexNet, GoogLeNet, and VGG. In contrast to fully connected networks, it appears that it is possible to exploit sparsity in CNNs, but it may be more challenging to get similar benefits. In particular, it is not clear how to apply this technique to very efficient CNN architectures, such as systolic arrays, since these are based around a regular dense matrix multiplication.

Finally, computer architects have also begun to consider sparsity in the model parameterization. This is especially relevant to fully connected layers, which often represent the vast majority of the total model size. Since there is no reuse in nonbatched fully connected layers, exploiting weight sparsity allows a significant reduction in the number of operations required. We

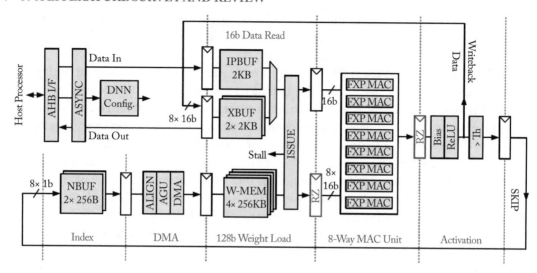

Figure 5.3: *DNN Engine* [144] microarchitecture exploits activation sparsity to reduce the number of cycles required to process a fully connected layer. A comparator is introduced in the activation stage that thresholds activation values. The resulting SKIP signal is used to predicate writeback for the following layer. A small SRAM (NBUF) is also introduced to store the indexes of the active neurons in order to generate weight addresses.

previously described some of the key approaches to model compression explored in the machine learning domain. Of these, an architecture has been developed for the *Deep Compression* [59] approach, called *EIE* [61]. The proposed architecture jointly considers both activation sparsity and weight sparsity. In general, this is achieved by sparse encoding both the weights and the activation data and using multiple decoupled processing elements to operate directly (in parallel) on this representation. Each element performs a single MAC operation and contains storage for the sparse indices. Since each decoupled element has a different number of operations to perform for each neuron in the network, there is a potential load-balancing problem, which is relaxed using FIFOs. One of the impressive results of this work is a near-linear speedup observed when scaling the number of processing elements. These results suggest that compression of fully connected layers may soon become an important technique for neural network hardware accelerators.

5.4.2   MODEL SUPPORT

Specialized hardware for neural network inference can quite readily achieve gains in performance and energy efficiency compared to off-the-shelf SIMD CPU or GPU solutions. However, architects are also cognizant that this comes with a trade-off in flexibility. The current situation in which neural networks are evolving very quickly presents a particular challenge, and architects should be prudent to carefully consider flexibility and future-proofing when designing specialized hardware. After all, custom hardware typically must be useful for many years to offset the exorbitant design and manufacturing costs. A domain-specific programmable architecture is the next step in terms of specialization without compromising programmability. However, research in this direction often raises the question of what kind of elementary operations should be supported. An instruction set architecture (ISA) that can achieve performance or efficiency superior to that of general purpose machines, while retaining some flexibility, would be a compelling result.

One of the first accelerator designs published that targets modern deep learning techniques was *DianNao* [19]. The accelerator architecture (Figure 5.4) is reminiscent of VLIW designs and consists of three scratchpad buffers, a datapath, and a control processor with a simple ISA. Buffering is provided for input activation data (NBin), output activation data (NBout), and weights (SB). Each buffer includes DMA logic. The pipelined datapath, called the *Neural Functional Unit* (NFU) implements multiple operations, such as multiply-accumulate, shifting, max, sigmoid, etc. The simulation results for a 65 nm implementation, running around 1 GHz, suggest that a simple programmable architecture based around this ISA can outperform commodity CPU (with SIMD) and GPU implementations to a great extent, with a huge average speedup of 118×. This is done while supporting the key layer types, including convolutional, pooling, and fully connected. Subsequent variations of this design [25, 45] address different application performance points, and exploit embedded DRAM on-chip to avoid costly off-chip memory accesses for large models.

While neural networks currently attract the most interest, other machine learning techniques may gain more popularity in the future. *PuDianNao* [97] is a recent development of previous DianNao accelerators that really explores the architecture flexibility question by including support for machine learning techniques such as k-means, k-nn, naive Bayes, Support Vector Machines, and linear regression. To achieve this, the datapath in particular becomes significantly more complex to support more operation types and requires 6 pipeline stages. Reported simulation results in 65 nm technology show that for seven diverse machine learning techniques, average speedup over a SIMD CPU is 21× and over a GPU is 1.2×. These results are significantly smaller than the original DianNao performance benefit, which may indicate the cost associated with increased model support.

Cambricon [98] is an ISA and implementation specifically targeting neural networks. The study proposes a load-store architecture with 64-bit instructions. A bank of 64 general purpose registers (32-bit) are used for scalars for control and addressing, with the remainder of the ISA focused on supporting intensive vector/matrix operations that form the basis of most neural network techniques. One significant difference here compared to a general purpose ISA is the use of a software managed scratchpad memory instead of a vector register file. The vector and matrix operands are stored in this scratchpad under full software control. In total, four instruction types are described (Figure 5.5): control, data transfer, computational, and logical. With this, it is pos-

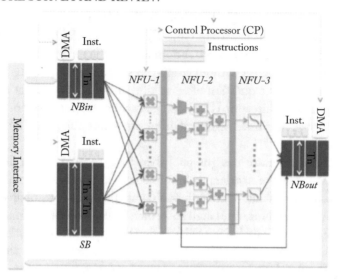

Figure 5.4: *DianNao* neural network accelerator architecture [19], consisting of three scratch-pad buffers (NBin, NBout, SB), a pipelined datapath (NFU), DMA per buffer, and a control processor (CP).

sible to efficiently support as many as ten benchmarks consisting of different machine learning techniques. Simulation results of a prototype accelerator implementing the *Cambricon* ISA in 65 nm technology show average speedup of about 92× compared to CPU and 3× compared to GPU implementations. This level of performance is broadly comparable to the DaDianNao accelerator, while offering much deeper model support. Therefore, this is a data point that suggests it is, in fact, possible to design an architecture that provides sufficient programmability for extensive model support while retaining enough specialism to achieve worthwhile performance gains.

The Google Tensor Processing Unit (TPU) [80], is an ASIC architecture that is essentially a matrix multiplication coprocessor for accelerating neural network inference in the datacenter. The proposed ASIC is used in production by the company in their datacenters. The paper itself provides a fascinating insight into the breakdown of the workloads that Google runs in their datacenters, which perhaps surprisingly, seem to strongly emphasize fully connected and LSTM layers, as opposed to CNNs. Figure 5.6 gives a block diagram for the TPU architecture, which is surprisingly simple. The heart of the architecture is a huge systolic matrix multiplication unit with sixty-four thousand 8-bit MACs. A large on-chip SRAM buffer is then used to feed the matrix unit, along with DDR3 DRAM ports that supply the weights. A PCIe interface is used to communicate with the host CPU that sends commands directly to the TPU. These commands consist of TPU instructions, which, in contrast to the previous architectures, are CISC-style and take an average of 10–20 clock cycles per instruction to complete. The key instructions are simple

Instruction Type		Examples	Operands
Control		Jump, conditional branch	Register (scalar value), immediate
Data Transfer	Matrix	Matrix load/store/move	Register (matrix address/size, scalar value), immediate
	Vector	Vector load/store/move	Register (vector address/size, scalar value), immediate
	Scalar	Scalar load/store/move	Register (scalar value), immediate
Computational	Matrix	Matrix multiply vector, vector multiply matrix, matrix multiply scalar, outer product, matrix add matrix, matrix subtract matrix	Register (matrix/vector address/ size, scalar value)
	Vector	Vector elementary arithmetics (add, subtract, multiply, divide), vector transcendental functions (exponential, logarithmic), dot product, random vector generator, maximum/minimum of a vector	Register (vector address/size, scalar value)
	Scalar	Scalar elementary arithmetics, scalar transcendental functions	Register (scalar value), immediate
Logical	Vector	Vector compare (greater than, equal), vector logical operations (and/or inverter), vector greater than merge	Register (vector address/size, scalar)
	Scalar	Scalar compare, scalar logical operations	Register (scalar), immediate

Figure 5.5: *Cambricon* [98] ISA consisting of four instruction types.

and fairly self-explanatory: Read_Host_Memory (into unified buffer), Write_Host_Memory (from unified buffer), Read_Weights (into weight FIFO), Matrix_Multiply, Convolve, and Activate (perform nonlinear activation function). Using these simple instructions, this architecture supports the key neural network layer types of fully connected, CNN and LSTM.

The TPU paper is a great case study for the potential benefits of a bespoke ASIC for datacenter inference applications. Average speedup for the TPU is 15× to 30× compared to contemporary CPU or GPU and achieves 30× to 50× higher TOPS/W. Characterization of CPU

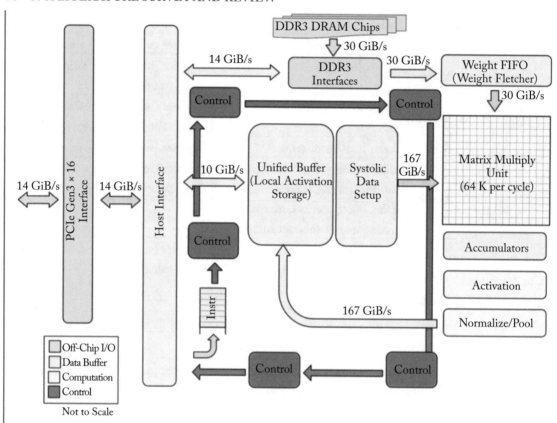

Figure 5.6: *TPU* [80] architecture, dominated by the large 64 k matrix multiply unit and unified buffer.

and GPU performance on the same fully connected, CNN, and LSTM workloads is shown in Figure 5.7, which presents this comparison in the form of an insightful roofline plot. This plot reminds us about some of the important characteristics of neural network layer types, i.e., fully connected and LSTM layers are clustered against the slanted part of the roofline, which indicates they are memory bound, whereas CNN layers are compute bound. For CPU and GPU, the neural network workloads are generally further below their rooflines, which is due to the strict response time that is demanded by datacenter applications. Comparing the CPU/GPU and TPU, the latter favors minimalism of a domain-specific processor rather than wasting energy and silicon area on unnecessary features such as caches, branch predictors, speculation, multi-threading, and so on, which generally accounts for the impressive performance and energy improvements. In terms of model support, TPU allows for acceleration of the two key operations of matrix multiplication and convolution.

Log-Log Scale

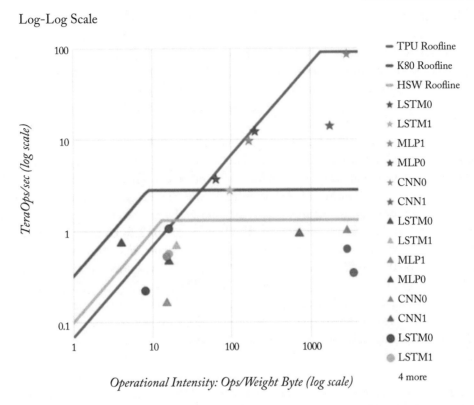

Figure 5.7: Roofline plot for Intel Haswell CPU (circles), Nvidia K80 GPU (triangles), and Google TPU (stars), for various MLP (fully connected), CNN and LSTM inference workloads.

We alluded earlier that FPGAs could be a very compelling platform on the performance/programmability continuum for accelerating machine learning workloads. Neural network inference accelerators implemented using the *Catapult* framework [17] make use of commercial off-the-shelf FPGAs in datacenter applications. The CNN inference accelerator architecture employs a systolic MAC array, which is a popular choice due to the efficiency of the local register-to-register data movement. Compared to CPU or GPU implementations, there are some advantages of predictable latency with FPGA, which is an important consideration in the datacenter. However, the performance results demonstrated to date seem fairly moderate, running CNNs at around 2.3× the throughput of a contemporary dual-socket server. This is significantly less than we saw with the TPU. It's also worth noting that potential future improvements with FPGA-based neural network accelerators is of course linked to the technology scaling of the underlying FPGA products themselves, which may start to falter as Moore's law begins to fade.

Although the performance and energy efficiency advantage of FPGA may be less than a dedicated ASIC, the development costs are also much lower. Having said that, FPGA development is not as easy as traditional software development. And, in fact, the learning curve involved in implementing a given workload on an FPGA platform is a potential roadblock in the wider adoption of FPGAs. This process is akin to ASIC design and typically involves low-level RTL design, which is slow and requires expertise in hardware design. *DNN Weaver* [124] addresses the design bottleneck of FPGAs with a parameterized inference accelerator that can be easily mapped to a given neural network architecture and FPGA platform. A dataflow graph derived from a neural network modeled in Caffe is used as the abstracted programming interface, which DNN Weaver converts to an accelerator using a parameterized Verilog design template and an FPGA specification. This automated approach minimizes the design effort, as long as the neural network is supported by the parameterized accelerator. Supported layer types are CNN and fully connected, but there is no mention of RNN support. The FPGA accelerators offer 1.3–7.1× average performance increase compared to a CPU baseline, but actually does worse than a GPU baseline by 0.15–0.85×. Power efficiency of the FPGA, however, is much better than CPU (8.7–15×) and GPU (2–3.6×). FPGAs are probably most likely to gain traction in datacenters, perhaps initially for inference applications in the cloud. However, performance results need to be much more competitive with commodity GPUs for this to take off.

The architecture work discussed thus far has been focused on accelerating inference. In contrast, very little work has been published to date on accelerating training with specialized architecture. *Tabla* [101] is one of the first architecture studies, around training, in this case using FPGAs. This paper follows a similar approach to DNN Weaver in providing a parameterized design that can be quickly targeted to a given neural network and FPGA platform. Stochastic gradient descent (SGD) is widely used for training of a variety of machine learning models, including linear regression, SVMs, and neural networks. Therefore, FPGA accelerators can be generated for a range of parameterizations of SGD using templating techniques. Results show average performance improvements of 19× and 3× compared to ARM A15 CPU and Intel Xeon CPU implementations respectively. In terms of energy efficiency, FPGA SGD accelerators offer improvements of 63× and 38× improvement in average performance-per-Watt for ARM A15 and Intel Xeon CPUs respectively. In contrast to inference, training is often a very hands-on process and it is not currently clear how things like debugging training issues would work when running SGD on an FPGA accelerator. And, as with DNN Weaver, the template approach is only as good as the templated design itself and its model support limitations.

In surveying architecture papers, one particular class of neural networks stands out as receiving by far the least attention: recurrent neural nets. RNNs have recently gained a lot of attention for results in natural language processing (NLP) applications, such as speech recognition, translation, sentiment analysis, and a host of other interesting uses. Nonetheless, we have seen very few papers that mention RNNs whatsoever. RNNs in their various flavors represent a much more complex dataflow graph than fully connected and CNN layers and perhaps have a less obvious mapping to hardware.

5.4.3 DATA MOVEMENT

The computation of neural networks involves working with a number of large vectors and matrices. As is often the case in computer architecture, the cost of manipulating and moving data can be more of a challenge than the actual computation itself. Since most neural networks have ample parallelism, memory bandwidth can rapidly become the limiting factor. Model compression and sparse encoding optimizations described previously can reduce the data movement and may allow some models to avoid costly DRAM access. However, the deep learning movement has always favored bigger models, hence they may inevitably need to spill over into main memory. Therefore, optimizing algorithms and architectures to minimize data movement is likely to be a significant area for future research.

A systolic array is an efficient approach to regular kernels such as matrix multiplication and convolution. The efficiency derives from the local movement of data to and from adjacent computational elements. This is typically register-to-register movement in a spatial pattern. A great example of this approach is the *TPU* [80], which uses a very large systolic array. However, such a regular structure is somewhat inflexible, and basically limited to a couple of operations.

Since neural networks exhibit extreme parallelism, it has been suggested that traditional SIMD/SIMT architectures may not scale well due to a bottleneck effect around structures such as register files and memory interfaces. Since much of the data movement in dataflow architectures is local, they may scale much more efficiently. Dataflow architectures employ many computation elements arranged in an interconnecting communication network. Operands flow through the network in a pattern derived from the underlying computation graph.

Neuflow [111] is a dataflow architecture for accelerating general purpose vision algorithms on FPGA. A compiler was also designed to help transform a high-level flow-graph representation of a neural network into machine code for the proposed architecture. Figure 5.8 shows the the dataflow architecture, which is a 2D grid of processing tiles (PTs), which process streams of data in parallel. A smart DMA unit interfaces the grid to off-chip memory. The operation of the system is determined using a runtime configuration controller and bus. The system targets real-time detection, categorization, and localization of objects in complex scenes, with measurement results demonstrating throughput of 12 frames per second on a 20-category problem.

Eyeriss [22] is also a spatial dataflow architecture, this time targeting CNN layers (Figure 5.9). An interesting aspect of dataflow machines is that there are often many ways to map a given computation graph to the spatial array. The CNNs computation typically has as many as seven degrees of freedom in scheduling the dataflow, and therefore there are many ways to map to a given dataflow architecture. In this work, a number of different dataflows are considered and classified based on how certain types of data move in a convolutional layer. From this analysis, a new dataflow is identified that minimizes energy consumption by maximizing input data

A Runtime Reconfigurable Dataflow Architecture

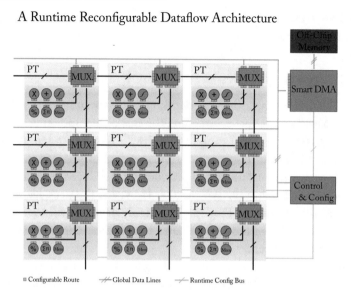

Figure 5.8: *Neuflow* [111] is a CNN accelerator system composed of a dataflow spatial accelerator and a DRAM.

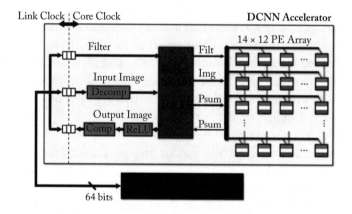

Figure 5.9: A CNN accelerator system composed of a dataflow spatial accelerator and a DRAM. [22]

reuse (i.e., weights and feature maps) and simultaneously minimizing partial sum accumulation cost. Compared to more obvious dataflows previously considered, this achieves improvements in energy of 1.4–2.5× for AlexNet.

In addition to optimizing the data sequencing within any given layer (intra-layer), the order in which the layers themselves are processed is also important (inter-layer) in terms of minimizing data movement. While most accelerators process CNN layers to completion, this often requires the use of off-chip memory to store intermediate data between layers. In [5], an accelerator is proposed that explores the dataflow across convolutional layers. In this way, the processing of multiple CNN layers can be *fused* by modifying the order in which the input data are brought on chip, such that inter-layer intermediate data between adjacent CNN layers can be cached. This work is especially exciting because it leverages the flexibility of hardware to freely schedule the workload. It's also one of the few studies to take a system-level view. Results show an FPGA CNN accelerator designed in this way can reduce total off-chip data transfers by 95% for the convolutional layers of VGG, reducing the data from 77 MB down to 3.6 MB per image.

Cutting-edge deep learning research is often still limited by computational resources. The time it takes to train a large model is a very real limitation on practical experiments. DRAM capacity on GPU hardware is also a fundamental limit. The many widely used machine learning frameworks require users to carefully tune memory usage to fit the model into the DRAM available on the GPU. Addressing this particular problem, *vDNN* [116] is a runtime memory manager that virtualizes the memory usage of DNNs. This means that both GPU and CPU memory can be simultaneously utilized to train larger models. For example, results show *vDNN* can reduce the average memory usage of AlexNet by up to 89%, with only minor performance loss.

5.5   CIRCUITS

An important historical circuits trend in computing neural networks is the interest in using analog circuits to implement datapath operations. Biological neural networks operate using an essentially analog and self-timed substrate, albeit a complex organic electrical-chemical system. It may appear that analog circuits would be an ideal candidate for implementing artificial neural networks. However, analog and mixed-signal techniques are encumbered with their own particular challenges, especially in regard to very practical issues relating to reliable mass manufacturing and test. Much of the interest in this has been focused on spiking neural networks, which are the class of models most closely related to biological networks. Spiking networks are not covered here; however, interest in analog approaches for deep learning has begun to grow. In this section, we survey some circuit techniques that may offer particular benefit to deep learning applications.

5.5.1   DATA MOVEMENT

We previously described the use of small data types in neural networks. The use of analog techniques are particularly well suited to situations that do not require very high precision, such as this. By replacing low-precision digital MAC operations with their analog counterpart, it may be possible to reduce the energy consumption of the calculation. *Redeye* [93] explores the use of analog circuits specifically for the application domain of computer vision. In particular, the observation that computer vision applications are often tolerant of low-SNR input images, and relatively low-SNR computation, motivates the use of analog circuits. The premise of this work is specifically to move data from early convolutional layers into the analog domain directly after the image sensor circuitry, before the ADC. Simulation results reported suggest that a reduction in sensor energy of 85× may be possible. Analog circuits certainly present challenges in terms of tolerance to variation effects and testability. To fully understand these challenges, real silicon prototyping is required to demonstrate a practical application of analog computing concepts.

ISAAC [122] proposes an accelerator with an in situ approach to dot-product operations, where memristor crossbar arrays are employed. One of the challenges of using a crossbar memory as an analog dot-product engine is the high overheads of analog-to-digital conversion. Therefore, special data-encoding techniques are developed that are more amenable to analog computations. Figure 5.10 illustrates the analog computation based around a resistive memory array. A series of input voltages are applied to the word lines, with the resistances of the memory cells preprogrammed. The voltage source in series with the resistance of the memory cell causes a scaled current to flow into the bit line for each input voltage, forming an analog sum of products operation. The basic memory array is then fleshed out with additional circuitry, such as DACs on the word-line inputs, and sample-and-hold and ADC circuits on the bit-line outputs. Simulation results at 32 nm on a suite of CNN and DNN workloads yield average improvements in throughput of 14.8× and reduction in energy of 5.5×, relative to the *DianNao* architecture.

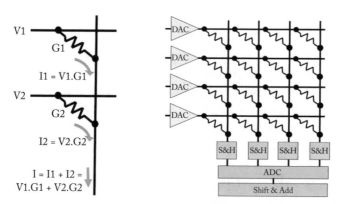

Figure 5.10: *ISAAC* [122] uses a memory-array bitline as a voltage input, in series with a resistive memory cell, which performs the multiplication. The resulting current flows into the bitline where it is accumulated with currents from the other cells in the column. Additional circuitry such as D/A and A/D converters are required to complete the functional unit (right).

Similarly, *PRIME* [27] considers another NVM technology: resistive random access memory (ReRAM). A ReRAM crossbar array is used to perform matrix-vector multiplication used in neural networks. Microarchitecture and circuit designs are presented that enable reconfigurable functionality with insignificant overhead. One of the critical challenges in analog crossbar computation is the precision requirement. There are essentially three aspects to this:

precision of input data to the crossbar, precision of the cell weight (resistance), and precision of the final output analog computation. All of these have an impact on the final predication accuracy of the neural network. In the case of ReRAM devices, current results suggest crossbar arrays may be able to support 7-bit output fidelity, which is certainly close to the range typically required for neural networks. The *PRIME* design employs 3-bit input precision (i.e., 8 voltage levels), 4-bit weight precision (16 resistance levels), and targets 6-bit output precision (64 levels). Simulation results presented suggest that performance improvements of 2,360× and energy reduction of 895× can be achieved across a set of machine learning benchmarks at the 65 nm process node.

In [12], the authors describe a ReRAM crossbar machine for calculating Boltzmann machine models. As with the previous examples, the computation is performed in parallel within the memory arrays and therefore eliminates the need to move data from the memory to the execution units. Three applications are studied in this context: graph partitioning, boolean satisfiability, and a deep belief network. Simulation results in a 22 nm technology show a performance increase of 57× and 25× lower energy with negligible loss in quality for these workloads, compared to a multicore CPU system.

In a similar vein to other analog approaches, resistive memory crossbar dot products are subject to many challenges in terms of process, voltage, and temperature (PVT) variation, as well as challenges with circuit noise and offset, testability, and the lack of robust analog memories. The architecture studies just described suggest great promise with analog crossbars, but we need to build real silicon implementations to understand how realistic simulation results are.

Neurocube [82] is based on a 3D stacked memory with a logic tier for processing neural networks, and considers both inference and training. More specifically, the architecture consists of clusters of processing engines connected by a 2D mesh network and integrated into a 3D stack with multiple tiers of DRAM (Figure 5.11). The 3D DRAM stack allows for terrific memory bandwidth, since it can be accessed in parallel, without the bottleneck of a conventional off-chip memory interface. Simulation results are presented for a 28 nm, 5 GHz implementation. A comparison of the 3D stacked solution with a more conventional DDR3 version highlights the advantages of the former. Although the peak bandwidth of DDR3 is higher than the 3D counterpart, DDR3 shows much lower system performance. This is because the DDR3 bandwidth is delivered over only two channels, which creates a bottleneck in the NoC that then must deliver data to the processing elements. The 3D DRAM interface, on the other hand, has much more parallelism and can deliver bandwidth spatially across the die.

5.5.2   FAULT TOLERANCE

It may seem apparent from the biological analogy that artificial neural network techniques would offer useful tolerance to faults in the underlying substrate. As one might expect in such a dense computational graph, failure of a small number of nodes in a neural network model often has little impact on classification accuracy. For computer architects, this particular quality is highly

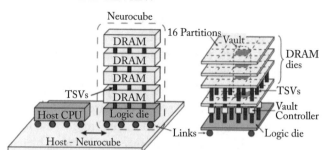

Figure 5.11: *Neurocube* [82] is based on a 3D stacked memory with DRAM tiers stacked on top of a logic tier, and targets neural networks.

prized. Designing, manufacturing, and testing a silicon chip that contains many billions of individual transistors is very challenging, and hence any opportunity to relax this "burden of correctness" is of great interest.

Indeed, there is evidence that neural networks exhibit fault tolerance properties that are useful in relaxing requirements for manufacturing faults in silicon chips. The study in [137] investigates the effect of injecting transistor-level faults in the logic gates of neural network hardware accelerators for simple fully connected DNNs. The empirical experiments in this study emphasize that maximizing parallelism increases the fault tolerance observed. The intuition behind this observation is that when we employ time multiplexing to fit a larger network onto a smaller set of hardware computation units, a single hardware fault can affect multiple nodes in the network. Although the considered networks and datasets are very small by deep learning standards, the brief results in this work suggest that a significant number of faults can be tolerated with negligible loss in classification accuracy.

It is also possible to trade error tolerance for efficiency or performance on fullyfunctional chips. In [44], the authors propose the use of inexact arithmetic, which allows for minor errors in the output computation in return for reduced logic complexity, potentially leading to lower power and latency in the evaluation of fully connected DNNs. The errors arising due to the inexact computations can then be compensated for during training, as illustrated in Figure 5.12. Simulation results in 65 nm technology show that this approach can achieve 1.78–2.67× reduction in energy, along with 1.23× delay reduction and 1.46× area saving, compared to the baseline.

The work in [87] explores a similar approach of introducing hardware approximations that reduce energy in fully connected DNNs. More specifically, an approach is presented to determine a set of weights that have little impact on output error. This is done by examining a sensitivity factor computed during backpropagation in the training process. The weights identified for approximation are then subject to reduced bit precision and approximate multiplier techniques. Simulation results in 130 nm technology show a 38% system power reduction with

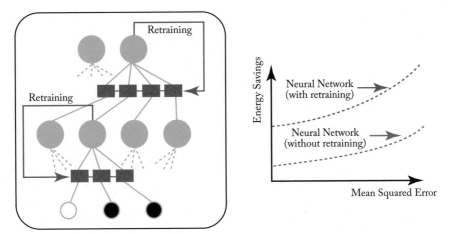

Figure 5.12: The ability of neural networks to train around circuit imperfections can be exploited to allow inexact computations, which are more efficient the conventional exact computations [44].

a 0.4% loss in classification accuracy on the MNIST dataset. Clearly the benefits that can be achieved at the circuit level are much more modest than we typically see higher up the stack.

An alternative approach is to use overclocking or undervolting of digital circuits to significantly increase throughput or reduce power at the cost of a small number of transient errors. The *DNN Engine* [144] is a fully connected DNN accelerator, demonstrated in silicon, that is robust to timing violations that intermittently occur when the supply voltage is aggressively scaled. This robustness is achieved through a combination of algorithmic and circuit-level timing-error tolerance techniques. The algorithmic error tolerance is demonstrated for both weights memory and datapath operations, showing the weights are very tolerant to errors, while the datapath exhibits relatively poor tolerance. Hence, the error tolerance of the datapath is additionally bolstered using circuit techniques originally developed for DSP accelerators in [143]. Measurement results of the 28 nm test chip demonstrate power savings of 30% from aggressive voltage scaling or, alternatively, a throughput improvement of 80% from aggressive clock frequency scaling. Both of these results are without any loss of accuracy on the entire MNIST test set, with aggregate error rates of greater than 10^{-1}. These results are compelling, but more studies on bigger models and datasets and other layer types are needed to better understand the limitations of fault tolerance in neural networks.

CHAPTER 6

Conclusion

Beyond a shadow of a doubt, deep learning is a protean field. The last decade has given rise to unprecedented progress in many fields through the use of practical deep learning techniques. Furthermore, the current level of interest from well-funded industry organizations is likely to ensure continued progress in the years to come.

Our experience of working on computer architecture for deep learning applications has bourne out that there are some key central concepts that have repeatedly provided fertile ground for optimizing the design of hardware support. A good example of one of these insights is the trade-off between the sparsity of neural network parameterizations, which seems to be central to the learning process, and the memory and compute requirements for inference. Like other central concepts described in this synthesis lecture, this is worthy of the attention of computer architects when applying our particular bag of tricks.

In particular, insights that arise from understanding and interacting directly with neural networks have often led us to the most useful observations. Therefore, we encourage computer architects to play with machine learning broadly rather than merely treating neural nets as a new "workload" to blindly profile and optimize. Perhaps this is to say that our concept of a workload needs to be inclusive of additions such as training and hyperparameter optimization, which would perhaps be locked down in an analog from conventional computer architecture workloads. We hope the workloads described in this synthesis lecture illustrate this. Indeed, there are often many degrees of freedom of the arrangement of connectionist models, probably much more so than in more traditional data processing algorithms. Therefore, there exists tremendous scope to co-optimize algorithm and architecture, which should be explored by computer architects.

Enabling researchers to freely explore this compelling, but often intimidating, space is perhaps the central focus of this work. There is no doubt that the emergence of excellent software tools for machine learning has been a significant boon. In turn, we need additional tools to allow us to explore the codesign space in terms of algorithms, architecture, and circuit techniques. Evolution implies that all three facets—models, data, and hardware—are still moving forward in lockstep. We look forward to the bright future of deep learning, and we believe that accurate, practical, and fluid workloads, tools, and computer architecture research will continue to play an important role in its progress.

Bibliography

[1] Martín Abadi, Ashish Agarwal, Paul Barham, Eugene Brevdo, Zhifeng Chen, Craig Citro, Greg S. Corrado, Andy Davis, Jeffrey Dean, Matthieu Devin, Sanjay Ghemawat, Ian Goodfellow, Andrew Harp, Geoffrey Irving, Michael Isard, Yangqing Jia, Rafal Jozefowicz, Lukasz Kaiser, Manjunath Kudlur, Josh Levenberg, Dan Mané, Rajat Monga, Sherry Moore, Derek Murray, Chris Olah, Mike Schuster, Jonathon Shlens, Benoit Steiner, Ilya Sutskever, Kunal Talwar, Paul Tucker, Vincent Vanhoucke, Vijay Vasudevan, Fernanda Viégas, Oriol Vinyals, Pete Warden, Martin Wattenberg, Martin Wicke, Yuan Yu, and Xiaoqiang Zheng. TensorFlow: Large-scale machine learning on heterogeneous distributed systems. *arXiv*, 1603.04467, 2016. http://arxiv.org/abs/1603.04467 30

[2] Robert Adolf, Saketh Rama, Brandon Reagen, Gu-Yeon Wei, and David M. Brooks. Fathom: Reference workloads for modern deep learning methods. In *IEEE International Symposium on Workload Characterization, (IISWC)*, pages 148–157, Providence, RI, IEEE Computer Society, September 25–27, 2016. https://doi.org/10.1109/IISWC.2016.7581275 DOI: 10.1109/iiswc.2016.7581275. xiii

[3] A. Agarwal, B. C. Paul, S. Mukhopadhyay, and K. Roy. Process variation in embedded memories: Failure analysis and variation aware architecture. *IEEE Journal of Solid-state Circuits*, 40(9):1804–1814, September 2005. DOI: 10.1109/jssc.2005.852159. 48, 56

[4] Jorge Albericio, Patrick Judd, Tayler Hetherington, Tor Aamodt, Natalie Enright Jerger, and Andreas Moshovos. Cnvlutin: Ineffectual-neuron-free deep neural network computing. In *Proc. of the 43rd International Symposium on Computer Architecture, (ISCA'16)*, pages 1–13, Piscataway, NJ, IEEE Press, 2016. https://doi.org/10.1109/ISCA.2016.11 DOI: 10.1109/isca.2016.11. 73

[5] M. Alwani, H. Chen, M. Ferdman, and P. Milder. Fused-layer CNN accelerators. In *49th Annual IEEE/ACM International Symposium on Microarchitecture (MICRO)*, pages 1–12, October 2016. DOI: 10.1109/micro.2016.7783725. 83

[6] S. Anwar, K. Hwang, and W. Sung. Fixed point optimization of deep convolutional neural networks for object recognition. In *IEEE International Conference on Acoustics, Speech and Signal Processing (ICASSP)*, pages 1131–1135, April 2015. DOI: 10.1109/icassp.2015.7178146. 66

[7] J. V. Arthur, P. A. Merolla, F. Akopyan, R. Alvarez, A. Cassidy, S. Chandra, S. K. Esser, N. Imam, W. Risk, D. B. D. Rubin, R. Manohar, and D. S. Modha. Building block of a programmable neuromorphic substrate: A digital neurosynaptic core. In *The International Joint Conference on Neural Networks (IJCNN)*, pages 1–8, June 2012. DOI: 10.1109/ijcnn.2012.6252637. 6

[8] Dzmitry Bahdanau, Kyunghyun Cho, and Yoshua Bengio. Neural machine translation by jointly learning to align and translate. In *International Conference on Learning Representations, (ICLR)*, 2015. 31, 38

[9] M. G. Bellemare, Y. Naddaf, J. Veness, and M. Bowling. The arcade learning environment: An evaluation platform for general agents. *Journal of Artificial Intelligence Research*, 47:253–279, June 2013. 34

[10] James Bergstra, Frédéric Bastien, Olivier Breuleux, Pascal Lamblin, Razvan Pascanu, Olivier Delalleau, Guillaume Desjardins, David Warde-Farley, Ian Goodfellow, Arnaud Bergeron, et al. Theano: Deep learning on GPUs with python. In *NIPS BigLearning Workshop*, 2011. 30

[11] Jock A. Blackard. *Comparison of Neural Networks and Discriminant Analysis in Predicting Forest Cover Types*. Ph.D. thesis, Colorado State University, Fort Collins, CO, 1998. AAI9921979. 60

[12] M. N. Bojnordi and E. Ipek. Memristive boltzmann machine: A hardware accelerator for combinatorial optimization and deep learning. In *IEEE International Symposium on High Performance Computer Architecture (HPCA)*, pages 1–13, March 2016. DOI: 10.1109/hpca.2016.7446049. 85

[13] K. A. Bowman, J. W. Tschanz, S. L. Lu, P. A. Aseron, M. M. Khellah, A. Raychowdhury, B. M. Geuskens, C. Tokunaga, C. B. Wilkerson, T. Karnik, and V. K. De. A 45 nm resilient microprocessor core for dynamic variation tolerance. *IEEE Journal of Solid-state Circuits*, 46(1):194–208, January 2011. DOI: 10.1109/jssc.2010.2089657. 58

[14] David Bull, Shidhartha Das, Karthik Shivashankar, Ganesh S. Dasika, Krisztian Flautner, and David Blaauw. A power-efficient 32 bit arm processor using timing-error detection and correction for transient-error tolerance and adaptation to PVT variation. *IEEE Journal of Solid-state Circuits*, 46(1):18–31, 2011. DOI: 10.1109/jssc.2011.2111230. 57, 58

[15] Chris Callison-Burch, Philipp Koehn, Christof Monz, and Josh Schroeder. Findings of the workshop on statistical machine translation. In *Proc. of the 4th Workshop on Statistical Machine Translation*, 2009. DOI: 10.3115/1626431.1626433.

[16] Ana Cardoso-Cachopo. *Improving Methods for Single-label Text Categorization*. Ph.D. thesis, Instituto Superior Tecnico, Universidade Tecnica de Lisboa, 2007. 60

[17] A. M. Caulfield, E. S. Chung, A. Putnam, H. Angepat, J. Fowers, M. Haselman, S. Heil, M. Humphrey, P. Kaur, J. Y. Kim, D. Lo, T. Massengill, K. Ovtcharov, M. Papamichael, L. Woods, S. Lanka, D. Chiou, and D. Burger. A cloud-scale acceleration architecture. In *49th Annual IEEE/ACM International Symposium on Microarchitecture (MICRO)*, pages 1–13, October 2016. DOI: 10.1109/micro.2016.7783710. 71, 79

[18] Tianshi Chen, Zidong Du, Ninghui Sun, Jia Wang, Chengyong Wu, Yunji Chen, and Olivier Temam. DianNao: A small-footprint high-throughput accelerator for ubiquitous machine-learning. In *Proc. of the 19th International Conference on Architectural Support for Programming Languages and Operating Systems, (ASPLOS)*, 2014. DOI: 10.1145/2541940.2541967. 6, 54

[19] Tianshi Chen, Zidong Du, Ninghui Sun, Jia Wang, Chengyong Wu, Yunji Chen, and Olivier Temam. Diannao: A small-footprint high-throughput accelerator for ubiquitous machine-learning. In *Proc. of the 19th International Conference on Architectural Support for Programming Languages and Operating Systems, (ASPLOS'14)*, pages 269–284, New York, NY, ACM, 2014. http://doi.acm.org.ezp-prod1.hul.harvard.edu/10.1145/2541940.2541967 DOI: 10.1145/2541940.2541967. 75, 76

[20] Wenlin Chen, James T. Wilson, Stephen Tyree, Kilian Q. Weinberger, and Yixin Chen. Compressing convolutional neural networks. *CoRR*, abs/1506.04449, 2015. http://arxiv.org/abs/1506.04449 68

[21] Wenlin Chen, James T. Wilson, Stephen Tyree, Kilian Q. Weinberger, and Yixin Chen. Compressing neural networks with the hashing trick. *CoRR*, abs/1504.04788, 2015. http://arxiv.org/abs/1504.04788 68, 69

[22] Yu-Hsin Chen, Joel Emer, and Vivienne Sze. Eyeriss: A spatial architecture for energy-efficient dataflow for convolutional neural networks. In *Proc. of the 43rd International Symposium on Computer Architecture, (ISCA'16)*, pages 367–379, Piscataway, NJ, IEEE Press, 2016. https://doi.org/10.1109/ISCA.2016.40 DOI: 10.1109/isca.2016.40. 72, 73, 81, 82

[23] Yu-Hsin Chen, Tushar Krishna, Joel Emer, and Vivienne Sze. Eyeriss: An energy-efficient reconfigurable accelerator for deep convolutional neural networks. In *International Solid-state Circuits Conference, (ISSCC)*, 2016. DOI: 10.1109/isscc.2016.7418007. 52

[24] Yunji Chen, Tao Luo, Shaoli Liu, Shijin Zhang, Liqiang He, Jia Wang, Ling Li, Tianshi Chen, Zhiwei Xu, Ninghui Sun, and O. Temam. DaDianNao: A machine-learning

supercomputer. In *Proc. of the 47th International Symposium on Microarchitecture*, 2014. DOI: 10.1109/micro.2014.58. 6

[25] Yunji Chen, Tao Luo, Shaoli Liu, Shijin Zhang, Liqiang He, Jia Wang, Ling Li, Tian-shi Chen, Zhiwei Xu, Ninghui Sun, and Olivier Temam. Dadiannao: A machine-learning supercomputer. In *Proc. of the 47th Annual IEEE/ACM International Symposium on Microarchitecture, (MICRO-47)*, pages 609–622, Washington, DC, IEEE Computer Society, 2014. http://dx.doi.org.ezp-prod1.hul.harvard.edu/10.1109/MICRO.2014.58 DOI: 10.1109/micro.2014.58. 73, 75

[26] Sharan Chetlur, Cliff Woolley, Philippe Vandermersch, Jonathan Cohen, John Tran, Bryan Catanzaro, and Evan Shelhamer. cuDNN: Efficient primitives for deep learning. *arXiv*, 1410.0759, 2014. http://arxiv.org/abs/1410.0759 31

[27] P. Chi, S. Li, C. Xu, T. Zhang, J. Zhao, Y. Liu, Y. Wang, and Y. Xie. Prime: A novel processing-in-memory architecture for neural network computation in reram-based main memory. In *ACM/IEEE 43rd Annual International Symposium on Computer Architecture (ISCA)*, pages 27–39, June 2016. DOI: 10.1109/isca.2016.13. 84

[28] Kyunghyun Cho, Bart Van Merriënboer, Caglar Gulcehre, Dzmitry Bahdanau, Fethi Bougares, Holger Schwenk, and Yoshua Bengio. Learning phrase representations using RNN encoder-decoder for statistical machine translation. *arXiv*, 1406.1078, 2014. http://arxiv.org/abs/1406.1078 DOI: 10.3115/v1/d14-1179. 27

[29] François Chollet. Keras: Theano-based deep learning library. https://keras.io, 2015. 31, 47

[30] Dan Claudiu Cireşan, Ueli Meier, Luca Maria Gambardella, and Jürgen Schmidhuber. Deep, big, simple neural nets for handwritten digit recognition. *Neural Computation*, 22(12):3207–3220, December 2010. http://dx.doi.org/10.1162/NECO_a_00052 DOI: 10.1162/NECO_a_00052. 6

[31] Dan C. Ciresan, Ueli Meier, Jonathan Masci, Luca Maria Gambardella, and Jürgen Schmidhuber. High-performance neural networks for visual object classification. *CoRR*, abs/1102.0183, 2011. http://arxiv.org/abs/1102.0183 70

[32] D. C. Ciresan, U. Meier, L. M. Gambardella, and J. Schmidhuber. Convolutional neural network committees for handwritten character classification. In *Document Analysis and Recognition (ICDAR), International Conference on*, pages 1135–1139, September 2011. DOI: 10.1109/icdar.2011.229. 6

[33] Ronan Collobert, Samy Bengio, and Johnny Mariéthoz. Torch: A modular machine learning software library. Technical report, Idiap Research Institute, 2002. 30

[34] Matthieu Courbariaux and Yoshua Bengio. Binarynet: Training deep neural networks with weights and activations constrained to +1 or −1. *CoRR.* http://arxiv.org/abs/1602.02830 44, 66

[35] Matthieu Courbariaux, Yoshua Bengio, and Jean-Pierre David. Binaryconnect: Training deep neural networks with binary weights during propagations. *CoRR,* abs/1511.00363, 2015. http://arxiv.org/abs/1511.00363 66

[36] Mark Craven, Dan DiPasquo, Dayne Freitag, Andrew McCallum, Tom Mitchell, Kamal Nigam, and Seán Slattery. Learning to extract symbolic knowledge from the world wide web. In *Proc. of the 15th National/10th Conference on Artificial Intelligence/Innovative Applications of Artificial Intelligence, (AAAI'98)/(IAAI'98),* pages 509–516, 1998. http://dl.acm.org/citation.cfm?id=295240.295725 60

[37] George Cybenko. Approximation by superpositions of a sigmoidal function. *Mathematics of Control, Signals, and Systems,* 2(4):303–314, 1989. DOI: 10.1007/bf02134016. 14

[38] S. Das, D. Roberts, Seokwoo Lee, S. Pant, D. Blaauw, T. Austin, K. Flautner, and T. Mudge. A self-tuning DVS processor using delay-error detection and correction. *IEEE Journal of Solid-state Circuits,* 41(4):792–804, April 2006. DOI: 10.1109/jssc.2006.870912. 57

[39] S. Das, C. Tokunaga, S. Pant, Wei-Hsiang Ma, S. Kalaiselvan, K. Lai, D. M. Bull, and D. T. Blaauw. Razorii: In situ error detection and correction for PVT and ser tolerance. *IEEE Journal of Solid-state Circuits,* 44(1):32–48, January 2009. DOI: 10.1109/jssc.2008.2007145. 58

[40] Saumitro Dasgupta. Caffe to TensorFlow. https://github.com/ethereon/caffe-tensorflow 31

[41] Jeffrey Dean, Greg S. Corrado, Rajat Monga, Kai Chen, Matthieu Devin, Quoc V. Le, Mark Z. Mao, MarcÁurelio Ranzato, Andrew Senior, Paul Tucker, Ke Yang, and Andrew Y. Ng. Large scale distributed deep networks. In *Advances in Neural Information Processing Systems, (NIPS),* 2012. 30

[42] J. Deng, W. Dong, R. Socher, L.-J. Li, K. Li, and L. Fei-Fei. ImageNet: A large-scale hierarchical image database. In *Proc. of the Conference on Computer Vision and Pattern Recognition, (CVPR),* 2009. DOI: 10.1109/cvprw.2009.5206848.

[43] Misha Denil, Babak Shakibi, Laurent Dinh, Marc'Aurelio Ranzato, and Nando de Freitas. Predicting parameters in deep learning. *CoRR,* abs/1306.0543, 2013. http://arxiv.org/abs/1306.0543 70

[44] Z. Du, A. Lingamneni, Y. Chen, K. V. Palem, O. Temam, and C. Wu. Leveraging the error resilience of neural networks for designing highly energy efficient accelerators. *IEEE Transactions on Computer-aided Design of Integrated Circuits and Systems*, 34(8):1223–1235, August 2015. DOI: 10.1109/tcad.2015.2419628. 86, 87

[45] Zidong Du, Robert Fasthuber, Tianshi Chen, Paolo Ienne, Ling Li, Tao Luo, Xiaobing Feng, Yunji Chen, and Olivier Temam. Shidiannao: Shifting vision processing closer to the sensor. In *Proc. of the 42nd Annual International Symposium on Computer Architecture, (ISCA'15)*, pages 92–104, New York, NY, ACM, 2015. http://doi.acm.org.ezp-prod1.hul.harvard.edu/10.1145/2749469.2750389 DOI: 10.1145/2749469.2750389. 75

[46] Steve K. Esser, Alexander Andreopoulos, Rathinakumar Appuswamy, Pallab Datta, Davis Barch, Arnon Amir, John Arthur, Andrew Cassidy, Myron Flickner, Paul Merolla, Shyamal Chandra, Nicola Basilico Stefano Carpin, Tom Zimmerman, Frank Zee, Rodrigo Alvarez-Icaza, Jeffrey A. Kusnitz, Theodore M. Wong, William P. Risk, Emmett McQuinn, Tapan K. Nayak, Raghavendra Singh, and Dharmendra S. Modha. Cognitive computing systems: Algorithms and applications for networks of neurosynaptic cores. In *The International Joint Conference on Neural Networks (IJCNN)*, 2013. DOI: 10.1109/ijcnn.2013.6706746. 6

[47] Clément Farabet, Berin Martini, Polina Akselrod, Selçuk Talay, Yann LeCun, and Eugenio Culurciello. Hardware accelerated convolutional neural networks for synthetic vision systems. In *ISCAS*, pages 257–260, IEEE, 2010. DOI: 10.1109/iscas.2010.5537908. 6

[48] Kunihiko Fukushima. Neocognitron: A self-organizing neural network model for a mechanism of pattern recognition unaffected by shift in position. *Biological Cybernetics*, 36(4):193–202, 1980. DOI: 10.1007/bf00344251. 3, 25

[49] Ken-Ichi Funahashi. On the approximate realization of continuous mappings by neural networks. *Neural Networks*, 2(3):183–192, 1989. DOI: 10.1016/0893-6080(89)90003-8. 14

[50] John S. Garofolo, Lori F. Lamel, William M. Fisher, Jonathan G. Fiscus, and David S. Pallett. TIMIT acoustic-phonetic continuous speech corpus. *LDC93S1*, 1993. https://catalog.ldc.upenn.edu/LDC93S1 30, 33

[51] Leon A. Gatys, Alexander S. Ecker, and Matthias Bethge. Image style transfer using convolutional neural networks. In *Proc. of the IEEE Conference on Computer Vision and Pattern Recognition*, pages 2414–2423, 2016. DOI: 10.1109/cvpr.2016.265. 16

[52] Xavier Glorot and Yoshua Bengio. Understanding the difficulty of training deep feedforward neural networks. In *Proc. of the 13th International Conference on Artificial Intelligence and Statistics*, 2010. 18, 22

[53] Xavier Glorot, Antoine Bordes, and Yoshua Bengio. Deep sparse rectifier neural networks. In *International Conference on Artificial Intelligence and Statistics*, pages 315–323, 2011. 23, 56

[54] Alex Graves, Greg Wayne, and Ivo Danihelka. Neural turing machines. *arXiv*, 1410.5401. `https://arxiv.org/abs/1410.5401` 28

[55] Alex Graves, Santiago Fernández, Faustino Gomez, and Jürgen Schmidhuber. Connectionist temporal classification: Labelling unsegmented sequence data with recurrent neural networks. In *Proc. of the 23rd International Conference on Machine Learning, (ICML)*, 2006. DOI: 10.1145/1143844.1143891. 33

[56] R. Grosse. Which research results will generalize? `https://hips.seas.harvard.ed u/blog/2014/09/02/which-research-results-will-generalize`, 2014. 5

[57] Suyog Gupta, Ankur Agrawal, Kailash Gopalakrishnan, and Pritish Narayanan. Deep learning with limited numerical precision. In *Proc. of the 32nd International Conference on Machine Learning*, pages 1737–1746, 2015. 6

[58] Suyog Gupta, Ankur Agrawal, Kailash Gopalakrishnan, and Pritish Narayanan. Deep learning with limited numerical precision. *CoRR*, abs/1502.02551, 2015. `http://arxi v.org/abs/1502.02551` 67

[59] Song Han, Huizi Mao, and William J. Dally. Deep compression: Compressing deep neural network with pruning, trained quantization and huffman coding. *CoRR*, abs/1510.00149, 2015. `http://arxiv.org/abs/1510.00149` 70, 74

[60] Song Han, Jeff Pool, John Tran, and William J. Dally. Learning both weights and connections for efficient neural networks. *CoRR*, abs/1506.02626, 2015. `http://arxiv.or g/abs/1506.02626` 70

[61] Song Han, Xingyu Liu, Huizi Mao, Jing Pu, Ardavan Pedram, Mark Horowitz, and William Dally. EIE: Efficient inference engine on compressed deep neural network. In *Proc. of the 43rd International Symposium on Computer Architecture, (ISCA)*, 2016. DOI: 10.1109/isca.2016.30. 43, 74

[62] Awni Hannun, Carl Case, Jared Casper, Bryan Catanzaro, Greg Diamos, Erich Elsen, Ryan Prenger, Sanjeev Satheesh, Shubho Sengupta, Adam Coates, and Andrew Y. Ng. Deep speech: Scaling up end-to-end speech recognition. *arXiv*, 1412.5567, 2015. `http: //arxiv.org/abs/1412.5567` 33, 38

[63] Johann Hauswald, Yiping Kang, Michael A. Laurenzano, Quan Chen, Cheng Li, Trevor Mudge, Ronald G. Dreslinski, Jason Mars, and Lingjia Tang. Djinn and tonic: DNN as a service and its implications for future warehouse scale computers. In

Proc. of the 42nd Annual International Symposium on Computer Architecture, (ISCA'15), pages 27–40, ACM, 2015. http://doi.acm.org/10.1145/2749469.2749472 DOI: 10.1145/2749469.2749472. 6

[64] Kaiming He, Xiangyu Zhang, Shaoqing Ren, and Jian Sun. Deep residual learning for image recognition. *arXiv*, 1512.03385, 2015. http://arxiv.org/abs/1512.03385 DOI: 10.1109/cvpr.2016.90. 26, 33

[65] Kaiming He, Xiangyu Zhang, Shaoqing Ren, and Jian Sun. Delving deep into rectifiers: Surpassing human-level performance on imagenet classification. In *Proc. of the International Conference on Computer Vision*, 2015. DOI: 10.1109/iccv.2015.123. 18, 22

[66] Kaiming He, Xiangyu Zhang, Shaoqing Ren, and Jian Sun. Identity mappings in deep residual networks. In *European Conference on Computer Vision*, pages 630–645, 2016. DOI: 10.1007/978-3-319-46493-0_38. 27

[67] José Miguel Hernández-Lobato, Michael A. Gelbart, Brandon Reagen, Robert Adolf, Daniel Hernández-Lobato, Paul N. Whatmough, David Brooks, Gu-Yeon Wei, and Ryan P. Adams. Designing neural network hardware accelerators with decoupled objective evaluations. In *NIPS Workshop on Bayesian Optimization*, 2016. 49, 61

[68] Geoffrey E. Hinton and Ruslan R. Salakhutdinov. Reducing the dimensionality of data with neural networks. *Science*, 313(5786):504–507, 2006. DOI: 10.1126/science.1127647. 33

[69] Geoffrey E. Hinton, Nitish Srivastava, Alex Krizhevsky, Ilya Sutskever, and Ruslan Salakhutdinov. Improving neural networks by preventing co-adaptation of feature detectors. *CoRR*, abs/1207.0580, 2012. http://arxiv.org/abs/1207.0580 49, 67

[70] Geoffrey E. Hinton, Oriol Vinyals, and Jeffrey Dean. Distilling the knowledge in a neural network. *CoRR*, abs/1503.02531, 2015. http://arxiv.org/abs/1503.02531 70

[71] Sepp Hochreiter. *Untersuchungen zu Dynamischen Neuronalen Netzen*. Diploma thesis, Technische Universität München, 1991. 22

[72] Sepp Hochreiter and Jürgen Schmidhuber. Long short-term memory. *Neural Computation*, 9(8):1735–1780, 1997. DOI: 10.1162/neco.1997.9.8.1735. 27

[73] Alan L. Hodgkin and Andrew F. Huxley. A quantitative description of membrane current and its application to conduction and excitation in nerve. *The Journal of Physiology*, 117(4):500, 1952. DOI: 10.1113/jphysiol.1952.sp004764. 10

[74] Kurt Hornik, Maxwell Stinchcombe, and Halbert White. Multilayer feedforward networks are universal approximators. *Neural Networks*, 2(5):359–366, 1989. DOI: 10.1016/0893-6080(89)90020-8. 14

[75] Forrest N. Iandola, Matthew W. Moskewicz, Khalid Ashraf, Song Han, William J. Dally, and Kurt Keutzer. Squeezenet: Alexnet-level accuracy with 50x fewer parameters and <1 mb model size. *CoRR*, abs/1602.07360, 2016. http://arxiv.org/abs/1602.07360 68, 69

[76] Sergey Ioffe and Christian Szegedy. Batch normalization: Accelerating deep network training by reducing internal covariate shift. *CoRR*, abs/1502.03167, 2015. http://arxiv.org/abs/1502.03167 23, 29, 67

[77] S. M. Jahinuzzaman, J. S. Shah, D. J. Rennie, and M. Sachdev. Design and analysis of a 5.3-pj 64-kb gated ground SRAM with multiword ECC. *IEEE Journal of Solid-state Circuits*, 44(9):2543–2553, September 2009. DOI: 10.1109/jssc.2009.2021088. 57

[78] Yangqing Jia, Evan Shelhamer, Jeff Donahue, Sergey Karayev, Jonathan Long, Ross Girshick, Sergio Guadarrama, and Trevor Darrell. Caffe: Convolutional architecture for fast feature embedding. In *Proc. of the International Conference on Multimedia*, 2014. DOI: 10.1145/2647868.2654889. 30

[79] Thorsten Joachims. A probabilistic analysis of the rocchio algorithm with TFIDF for text categorization. In *ICML*, pages 143–151, 1997. http://dl.acm.org/citation.cfm?id=645526.657278 60

[80] Norman P. Jouppi, Cliff Young, Nishant Patil, David Patterson, Gaurav Agrawal, Raminder Bajwa, Sarah Bates, Suresh Bhatia, Nan Boden, Al Borchers, Rick Boyle, Pierre-luc Cantin, Clifford Chao, Chris Clark, Jeremy Coriell, Mike Daley, Matt Dau, Jeffrey Dean, Ben Gelb, Tara Vazir Ghaemmaghami, Rajendra Gottipati, William Gulland, Robert Hagmann, Richard C. Ho, Doug Hogberg, John Hu, Robert Hundt, Dan Hurt, Julian Ibarz, Aaron Jaffey, Alek Jaworski, Alexander Kaplan, Harshit Khaitan, Andy Koch, Naveen Kumar, Steve Lacy, James Laudon, James Law, Diemthu Le, Chris Leary, Zhuyuan Liu, Kyle Lucke, Alan Lundin, Gordon MacKean, Adriana Maggiore, Maire Mahony, Kieran Miller, Rahul Nagarajan, Ravi Narayanaswami, Ray Ni, Kathy Nix, Thomas Norrie, Mark Omernick, Narayana Penukonda, Andy Phelps, Jonathan Ross, Amir Salek, Emad Samadiani, Chris Severn, Gregory Sizikov, Matthew Snelham, Jed Souter, Dan Steinberg, Andy Swing, Mercedes Tan, Gregory Thorson, Bo Tian, Horia Toma, Erick Tuttle, Vijay Vasudevan, Richard Walter, Walter Wang, Eric Wilcox, and Doe Hyun Yoon. In-datacenter performance analysis of a tensor processing unit. *CoRR*, abs/1704.04760, 2017. http://arxiv.org/abs/1704.04760 DOI: 10.1145/3079856.3080246. 66, 70, 76, 78, 81

[81] Andrej Karpathy, Justin Johnson, and Li Fei-Fei. Visualizing and understanding recurrent networks. *arXiv*, 1506.02078, 2015. https://arxiv.org/abs/1506.02078 15

[82] D. Kim, J. Kung, S. Chai, S. Yalamanchili, and S. Mukhopadhyay. Neurocube: A programmable digital neuromorphic architecture with high-density 3D memory. In *ACM/IEEE 43rd Annual International Symposium on Computer Architecture (ISCA)*, pages 380–392, June 2016. DOI: 10.1109/isca.2016.41. 85, 86

[83] Jung Kuk Kim, Phil Knag, Thomas Chen, and Zhengya Zhang. A 640 m pixel/s 3.65 mw sparse event-driven neuromorphic object recognition processor with on-chip learning. In *Symposium on VLSI Circuits*, pages C50–C51, June 2015. DOI: 10.1109/vlsic.2015.7231323. 6

[84] Diederik P. Kingma and Max Welling. Stochastic gradient VB and the variational auto-encoder. In *2nd International Conference on Learning Representations, (ICLR)*, 2014. 33

[85] Alex Krizhevsky, Ilya Sutskever, and Geoffrey E. Hinton. ImageNet classification with deep convolutional neural networks. In *Advances in Neural Information Processing Systems, (NIPS)*, 2012. DOI: 10.1145/3065386. 26, 29, 34, 70

[86] J. P. Kulkarni, C. Tokunaga, P. Aseron, T. Nguyen, C. Augustine, J. Tschanz, and V. De. 4.7 a 409 gops/w adaptive and resilient domino register file in 22 nm tri-gate CMOS featuring in-situ timing margin and error detection for tolerance to within-die variation, voltage droop, temperature and aging. In *ISSCC*, pages 1–3, February 2015. DOI: 10.1109/isscc.2015.7062936. 57

[87] J. Kung, D. Kim, and S. Mukhopadhyay. A power-aware digital feedforward neural network platform with backpropagation driven approximate synapses. In *IEEE/ACM International Symposium on Low Power Electronics and Design (ISLPED)*, pages 85–90, July 2015. DOI: 10.1109/islped.2015.7273495. 86

[88] Jaeha Kung, Duckhwan Kim, and Saibal Mukhopadhyay. A power-aware digital feedforward neural network platform with backpropagation driven approximate synapses. In *ISPLED*, 2015. DOI: 10.1109/islped.2015.7273495. 6, 54

[89] Yann LeCun, Léon Bottou, Yoshua Bengio, and Patrick Haffner. Gradient-based learning applied to document recognition. *Proc. of the IEEE*, 86(11):2278–2324, 1998. DOI: 10.1109/5.726791. 3, 26

[90] Yann LeCun, Corinna Cortes, and Christopher J. C. Burges. The MNIST database of handwritten digits. http://yann.lecun.com/exdb/mnist/, 1998. 60

[91] Benjamin C. Lee and David M. Brooks. Illustrative design space studies with microarchitectural regression models. In *HPCA*, 2007. DOI: 10.1109/hpca.2007.346211. 61

[92] David D. Lewis.   Reuters-21578 text categorization collection data set. `https://archive.ics.uci.edu/ml/datasets/Reuters-21578+Text+Categor ization+Collection` 60

[93] R. LiKamWa, Y. Hou, Y. Gao, M. Polansky, and L. Zhong. Redeye: Analog convnet image sensor architecture for continuous mobile vision. In *ACM/IEEE 43rd Annual International Symposium on Computer Architecture (ISCA)*, pages 255–266, June 2016. DOI: 10.1109/isca.2016.31. 83

[94] Darryl Dexu Lin, Sachin S. Talathi, and V. Sreekanth Annapureddy. Fixed point quantization of deep convolutional networks. *CoRR*, abs/1511.06393, 2015. `http://arxiv.org/abs/1511.06393` 66

[95] Min Lin, Qiang Chen, and Shuicheng Yan. Network in network. *CoRR*, abs/1312.4400, 2013. `http://arxiv.org/abs/1312.4400` 68

[96] Zhouhan Lin, Matthieu Courbariaux, Roland Memisevic, and Yoshua Bengio. Neural networks with few multiplications. *CoRR*, abs/1510.03009, 2015. `http://arxiv.org/ abs/1510.03009` 66

[97] Daofu Liu, Tianshi Chen, Shaoli Liu, Jinhong Zhou, Shengyuan Zhou, Olivier Teman, Xiaobing Feng, Xuehai Zhou, and Yunji Chen. Pudiannao: A polyvalent machine learning accelerator. *SIGPLAN Notices*, 50(4):369–381, March 2015. `http://doi.acm.org.ezp-prod1.hul.harvard.edu/10.1145/2775054. 2694358` DOI: 10.1145/2775054.2694358. 75

[98] S. Liu, Z. Du, J. Tao, D. Han, T. Luo, Y. Xie, Y. Chen, and T. Chen. Cambricon: An instruction set architecture for neural networks. In *ACM/IEEE 43rd Annual International Symposium on Computer Architecture (ISCA)*, pages 393–405, June 2016. DOI: 10.1109/isca.2016.42. 75, 77

[99] Jean Livet, Tamily A. Weissman, Hyuno Kang, Ryan W. Draft, Ju Lu, Robyn A. nnis, Joshua R. Sanes, and Jeff W. Lichtman. Transgenic strategies for combinatorial expression of fluorescent proteins in the nervous system. *Nature*, 450(7166):56–62, November 2007. DOI: 10.1038/nature06293. 9

[100] David G. Lowe. Object recognition from local scale-invariant features. In *Proc. of the 7th IEEE International Conference on Computer Vision*, volume 2, pages 1150–1157, 1999. DOI: 10.1109/iccv.1999.790410. 25

[101] D. Mahajan, J. Park, E. Amaro, H. Sharma, A. Yazdanbakhsh, J. K. Kim, and H. Esmaeilzadeh. Tabla: A unified template-based framework for accelerating statistical machine learning. In *IEEE International Symposium on High Performance Computer Architecture (HPCA)*, pages 14–26, March 2016. DOI: 10.1109/hpca.2016.7446050. 80

[102] Jaakko Malmivuo and Robert Plonsey. *Bioelectromagnetism: Principles and Applications of Bioelectric and Biomagnetic Fields*. Oxford University Press, 1995. DOI: 10.1093/acprof:oso/9780195058239.001.0001. 10

[103] John McCarthy. Professor Sir James Lighthill, FRS. Artificial intelligence: A general survey. *Artificial Intelligence*, 5(3):317–322, 1974. http://dblp.uni-trier.de/db/journals/ai/ai5.html#McCarthy74 3

[104] Warren S. McCulloch and Walter Pitts. A logical calculus of the ideas immanent in nervous activity. *The Bulletin of Mathematical Biophysics*, 1943. DOI: 10.1007/bf02478259. 1

[105] Marvin Minsky and Seymour Papert. *Perceptrons: An Introduction to Computational Geometry*. MIT Press, 1969. 14

[106] Marvin Minsky and Seymour Papert. *Perceptrons: An Introduction to Computational Geometry*. MIT Press, Cambridge, MA, 1969. 2

[107] Volodymyr Mnih, Koray Kavukcuoglu, David Silver, Alex Graves, Ioannis Antonoglou, Daan Wierstra, and Martin Riedmiller. Playing atari with deep reinforcement learning. In *NIPS Deep Learning Workshop*, 2013. 34

[108] Volodymyr Mnih, Koray Kavukcuoglu, David Silver, Andrei A. Rusu, Joel Veness, Marc G. Bellemare, Alex Graves, Martin Riedmiller, Andreas K. Fidjeland, Georg Ostrovski, et al. Human-level control through deep reinforcement learning. *Nature*, 518(7540):529–533, 2015. DOI: 10.1038/nature14236. 34

[109] B. Moons and M. Verhelst. A 0.3–2.6 tops/w precision-scalable processor for real-time large-scale convnets. In *IEEE Symposium on VLSI Circuits (VLSI-Circuits)*, pages 1–2, June 2016. DOI: 10.1109/vlsic.2016.7573525. 72, 73

[110] Joachim Ott, Zhouhan Lin, Ying Zhang, Shih-Chii Liu, and Yoshua Bengio. Recurrent neural networks with limited numerical precision. *CoRR*, abs/1608.06902, 2016. http://arxiv.org/abs/1608.06902 67

[111] P. H. Pham, D. Jelaca, C. Farabet, B. Martini, Y. LeCun, and E. Culurciello. Neuflow: Dataflow vision processing system-on-a-chip. In *IEEE 55th International Midwest Symposium on Circuits and Systems (MWSCAS)*, pages 1044–1047, August 2012. DOI: 10.1109/mwscas.2012.6292202. 81, 82

[112] Christopher Poultney, Sumit Chopra, and Yann Lecun. Efficient learning of sparse representations with an energy-based model. In *Advances in Neural Information Processing Systems (NIPS6)*, MIT Press, 2006. 6

[113] A. Raychowdhury, B. M. Geuskens, K. A. Bowman, J. W. Tschanz, S. L. Lu, T. Karnik, M. M. Khellah, and V. K. De. Tunable replica bits for dynamic variation tolerance in 8 t SRAM arrays. *IEEE Journal of Solid-state Circuits*, 46(4):797–805, April 2011. DOI: 10.1109/jssc.2011.2108141. 57

[114] Brandon Reagen, Paul Whatmough, Robert Adolf, Saketh Rama, Hyunkwang Lee, Sae Kyu Lee, José Miguel Hernández-Lobato, Gu-Yeon Wei, and David Brooks. Minerva: Enabling low-power, highly-accurate deep neural network accelerators. In *Proc. of the 43rd International Symposium on Computer Architecture (ISCA)*, 2016. DOI: 10.1109/isca.2016.32. xiii, 43, 48, 72

[115] Brandon Reagen, José Miguel Hernández-Lobato, Robert Adolf, Michael A. Gelbart, Paul N. Whatmough, Gu-Yeon Wei, and David Brooks. A case for efficient accelerator design space exploration via Bayesian optimization. In *ISLPED*, 2017. 49, 61

[116] M. Rhu, N. Gimelshein, J. Clemons, A. Zulfiqar, and S. W. Keckler. VDNN: Virtualized deep neural networks for scalable, memory-efficient neural network design. In *49th Annual IEEE/ACM International Symposium on Microarchitecture (MICRO)*, pages 1–13, October 2016. DOI: 10.1109/micro.2016.7783721. 83

[117] Frank Rosenblatt. The perceptron, a perceiving and recognizing automaton: (project para). Technical report, Cornell Aeronautical Lab, 1957. 1

[118] Frank Rosenblatt. Principles of neurodynamics; perceptrons and the theory of brain mechanisms. Technical report, Cornell Aeronautical Lab, 1961. 12

[119] D. E. Rumelhart, G. E. Hinton, and R. J. Williams. Learning representations by back-propagating errors. *Nature*, 323:533–536, October 1986. DOI: 10.1038/323533a0. 2, 20

[120] Olga Russakovsky, Jia Deng, Hao Su, Jonathan Krause, Sanjeev Satheesh, Sean Ma, Zhiheng Huang, Andrej Karpathy, Aditya Khosla, Michael Bernstein, Alexander C. Berg, and Li Fei-Fei. ImageNet large scale visual recognition challenge. *International Journal of Computer Vision*, 115(3):211–252, 2015. DOI: 10.1007/s11263-015-0816-y. 26

[121] Adrian Sampson, Werner Dietl, Emily Fortuna, Danushen Gnanapragasam, Luis Ceze, and Dan Grossman. Enerj: Approximate data types for safe and general low-power computation. In *Proc. of the 32nd ACM SIGPLAN Conference on Programming Language Design and Implementation, (PLDI'11)*, pages 164–174, New York, NY, ACM, 2011. http://doi.acm.org/10.1145/1993498.1993518 DOI: 10.1145/1993498.1993518. 44

[122] Ali Shafiee, Anirban Nag, Naveen Muralimanohar, Rajeev Balasubramonian, J. Strachan, Miao Hu, R. Stanley Williams, and Vivek Srikumar. Isaac: A convolutional neural network accelerator with in-situ analog arithmetic in crossbars. *ISCA*, 2016. DOI: 10.1109/isca.2016.12. 84

[123] Yakun Sophia Shao, Brandon Reagen, Gu-Yeon Wei, and David Brooks. Aladdin: A pre-RTL, power-performance accelerator simulator enabling large design space exploration of customized architectures. In *Proc. of the 41st Annual International Symposium on Computer Architecuture, (ISCA'14)*, pages 97–108, Piscataway, NJ, IEEE Press, 2014. http://dl.acm.org/citation.cfm?id=2665671.2665689 DOI: 10.1109/isca.2014.6853196. 47

[124] H. Sharma, J. Park, D. Mahajan, E. Amaro, J. K. Kim, C. Shao, A. Mishra, and H. Esmaeilzadeh. From high-level deep neural models to FPGAS. In *49th Annual IEEE/ACM International Symposium on Microarchitecture (MICRO)*, pages 1–12, October 2016. DOI: 10.1109/micro.2016.7783720. 80

[125] Karen Simonyan and Andrew Zisserman. Very deep convolutional networks for large-scale image recognition. *arXiv*, 1409.1556, 2014. http://arxiv.org/abs/1409.1556 34

[126] Vikas Sindhwani, Tara N. Sainath, and Sanjiv Kumar. Structured transforms for small-footprint deep learning. *CoRR*, abs/1510.01722, 2015. http://arxiv.org/abs/1510.01722 68, 70

[127] Jasper Snoek, Hugo Larochelle, and Ryan P. Adams. Practical Bayesian optimization of machine learning algorithms. In F. Pereira, C. J. C. Burges, L. Bottou, and K. Q. Weinberger, Eds., *Advances in Neural Information Processing Systems 25*, pages 2951–2959, Curran Associates, Inc., 2012. http://papers.nips.cc/paper/4522-practical-bayesian-optimization-of-machine-learning-algorithms.pdf 61, 68

[128] Nitish Srivastava, Geoffrey Hinton, Alex Krizhevsky, Ilya Sutskever, and Ruslan Salakhutdinov. Dropout: A simple way to prevent neural networks from overfitting. *Journal of Machine Learning Research*, 15(1):1929–1958, January 2014. http://dl.acm.org/citation.cfm?id=2627435.2670313 6

[129] Rupesh Kumar Srivastava, Klaus Greff, and Jürgen Schmidhuber. Highway networks. *arXiv*, 1505.00387. https://arxiv.org/abs/1505.00387 27

[130] D. Strigl, K. Kofler, and S. Podlipnig. Performance and scalability of GPU-based convolutional neural networks. In *Euromicro Conference on Parallel, Distributed and Network-based Processing*, pages 317–324, February 2010. DOI: 10.1109/pdp.2010.43. 6

[131] Hendrik Strobelt, Sebastian Gehrmann, Bernd Huber, Hanspeter Pfister, and Alexander M. Rush. Visual analysis of hidden state dynamics in recurrent neural networks. *arXiv*, 1606.07461, 2016. https://arxiv.org/abs/1606.07461 15

[132] Evangelos Stromatias, Daniel Neil, Francesco Galluppi, Michael Pfeiffer, Shih-Chii Liu, and Steve Furber. Scalable energy-efficient, low-latency implementations of trained spiking deep belief networks on spinnaker. In *International Joint Conference on Neural Networks*, pages 1–8, IEEE, 2015. DOI: 10.1109/ijcnn.2015.7280625. 6

[133] Sainbayar Sukhbaatar, Arthur Szlam, Jason Weston, and Rob Fergus. End-to-end memory networks. In *Advances in Neural Information Processing Systems, (NIPS)*. 2015. http://papers.nips.cc/paper/5846-end-to-end-memory-networks.pdf 33

[134] Wonyong Sung, Sungho Shin, and Kyuyeon Hwang. Resiliency of deep neural networks under quantization. *CoRR*, abs/1511.06488, 2015. http://arxiv.org/abs/1511.06488 67

[135] Ilya Sutskever, Oriol Vinyals, and Quoc V. Le. Sequence to sequence learning with neural networks. In *Advances in Neural Information Processing Systems, (NIPS)*, 2014. 5, 31

[136] Christian Szegedy, Wei Liu, Yangqing Jia, Pierre Sermanet, Scott Reed, Dragomir Anguelov, Dumitru Erhan, Vincent Vanhoucke, and Andrew Rabinovich. Going deeper with convolutions. In *Proc. of the Conference on Computer Vision and Pattern Recognition, (CVPR)*, 2015. DOI: 10.1109/cvpr.2015.7298594. 26

[137] Olivier Temam. A defect-tolerant accelerator for emerging high-performance applications. In *Proc. of the 39th Annual International Symposium on Computer Architecture, (ISCA'12)*, pages 356–367, Washington, DC, IEEE Computer Society, 2012. http://dl.acm.org/citation.cfm?id=2337159.2337200 DOI: 10.1109/isca.2012.6237031. 86

[138] Giulio Srubek Tomassy, Daniel R. Berger, Hsu-Hsin Chen, Narayanan Kasthuri, Kenneth J. Hayworth, Alessandro Vercelli, H. Sebastian Seung, Jeff W. Lichtman, and Paola Arlotta. Distinct profiles of myelin distribution along single axons of pyramidal neurons in the neocortex. *Science*, 344(6181):319–24, 2014. DOI: 10.1126/science.1249766. 10

[139] Li Wan, Matthew Zeiler, Sixin Zhang, Yann LeCun, and Rob Fergus. Regularization of neural networks using dropconnect. In *International Conference on Machine learning*, 2013. 6

[140] P. J. Werbos. *Beyond Regression: New Tools for Prediction and Analysis in the Behavioral Sciences*. Ph.D. thesis, Harvard University, 1974. 2, 20

[141] Jason Weston, Sumit Chopra, and Antoine Bordes. Memory networks. *arXiv*, 1410.3916, 2014. `http://arxiv.org/abs/1410.3916` 28, 33

[142] Jason Weston, Antoine Bordes, Sumit Chopra, Alexander M. Rush, Bart van Merriënboer, Armand Joulin, and Tomas Mikolov. Towards AI-complete question answering: A set of prerequisite toy tasks. *arXiv*, 1502.05698, 2015. `http://arxiv.org/abs/1502.05698` 33

[143] P. N. Whatmough, S. Das, and D. M. Bull. A low-power 1-ghz razor fir accelerator with time-borrow tracking pipeline and approximate error correction in 65-nm CMOS. *IEEE Journal of Solid-state Circuits*, 49(1):84–94, January 2014. DOI: 10.1109/isscc.2013.6487800. 58, 87

[144] P. N. Whatmough, S. K. Lee, H. Lee, S. Rama, D. Brooks, and G. Y. Wei. 14.3 a 28 nm soc with a 1.2 ghz 568 nj/prediction sparse deep-neural-network engine with > 0.1 timing error rate tolerance for IOT applications. In *IEEE International Solid-state Circuits Conference (ISSCC)*, pages 242–243, February 2017. DOI: 10.1109/isscc.2017.7870351. 73, 74, 87

[145] Sergey Zagoruyko. loadcaffe. `https://github.com/szagoruyko/loadcaffe` 31

[146] Matthew D. Zeiler and Rob Fergus. Visualizing and understanding convolutional networks. In *European Conference on Computer Vision*, pages 818–833, 2014. DOI: 10.1007/978-3-319-10590-1_53. 15

Authors' Biographies

BRANDON REAGEN

Brandon Reagen is a Ph.D. candidate at Harvard University. He received his B.S. degree in Computer Systems Engineering and Applied Mathematics from University of Massachusetts, Amherst in 2012 and his M.S. in Computer Science from Harvard in 2014. His research spans the fields of Computer Architecture, VLSI, and Machine Learning with specific interest in designing extremely efficient hardware to enable ubiquitous deployment of Machine Learning models across all compute platforms.

ROBERT ADOLF

Robert Adolf is a Ph.D. candidate in computer architecture at Harvard University. After earning a B.S. in Computer Science from Northwestern University in 2005, he spent four years doing benchmarking and performance analysis of supercomputers at the Department of Defense. In 2009, he joined Pacific Northwest National Laboratory as a research scientist, where he lead a team building large-scale graph analytics on massively multithreaded architectures. His research interests revolve around modeling, analysis, and optimization techniques for high-performance software, with a current focus on deep learning algorithms. His philosophy is that the combination of statistical methods, code analysis, and domain knowledge leads to better tools for understanding and building fast systems.

PAUL WHATMOUGH

Paul Whatmough leads research on computer architecture for Machine Learning at ARM Research, Boston, MA. He is also an Associate in the School of Engineering and Applied Science at Harvard University. Dr. Whatmough received the B.Eng. degree (with first class Honors) from the University of Lancaster, U.K., M.Sc. degree (with distinction) from the University of Bristol, U.K., and Doctorate degree from University College London, U.K. His research interests span algorithms, computer architecture, and circuits. He has previously led various projects on hardware accelerators, Machine Learning, SoC architecture, Digital Signal Processing (DSP), variation tolerance, and supply voltage noise.

GU-YEON WEI

Gu-Yeon Wei is Gordon McKay Professor of Electrical Engineering and Computer Science in the School of Engineering and Applied Sciences (SEAS) at Harvard University. He received his B.S., M.S., and Ph.D. degrees in Electrical Engineering from Stanford University in 1994, 1997, and 2001, respectively. His research interests span multiple layers of a computing system: mixed-signal integrated circuits, computer architecture, and design tools for efficient hardware. His research efforts focus on identifying synergistic opportunities across these layers to develop energy-efficient solutions for a broad range of systems from flapping-wing microrobots to machine learning hardware for IoT/edge devices to specialized accelerators for large-scale servers.

DAVID BROOKS

David Brooks is the Haley Family Professor of Computer Science in the School of Engineering and Applied Sciences at Harvard University. Prior to joining Harvard, he was a research staff member at IBM T. J. Watson Research Center. Prof. Brooks received his B.S. in Electrical Engineering at the University of Southern California and M.A. and Ph.D. degrees in Electrical Engineering at Princeton University. His research interests include resilient and power-efficient computer hardware and software design for high-performance and embedded systems. Prof. Brooks is a Fellow of the IEEE and has received several honors and awards including the ACM Maurice Wilkes Award, ISCA Influential Paper Award, NSF CAREER award, IBM Faculty Partnership Award, and DARPA Young Faculty Award.